PRACTICAL PHOTOVOLTAICS
Electricity from Solar Cells

Richard J. Komp, PhD
Skyheat Associates
English, Indiana

Second Edition

FOREWORD

For over a century people have known that sunlight can produce electricity. The French physicist Edmond Becquerel discovered this fact in 1839. Like so many of his scientific contemporaries, Becquerel devoted much study to electricity. As part of a series of electrical experiments, he immersed two metal plates in a conductive fluid and exposed the apparatus to the sun. He observed a small voltage.

In another set of experiments, Willoughby Smith discovered, in 1873, that selenium was sensitive to light. Smith's discovery stimulated Adams and Day to conduct further tests with selenium and light, proving that when light struck selenium, an electrical current was generated.

About the same time that large electrical generating plants began producing power for factories, Charles Fritts came up with the first selenium solar cell. According to Fritts, when exposed to light the cell generated "a current that was continuous, constant, and of considerable electromotive force." Fritts had ambitious plans for his cells. He predicted that they "may ere long compete with the dynamo-electric machine." Due to the compactness of the cells, he felt that each building could be its own electrical generating plant rather than be dependent on a centralized grid. Furthermore, Fritts wrote in 1886 that since "the supply of solar energy is both without limit and without cost," it will continue to pour down upon us long after we run out of fossil fuels.

The nascent solar cell industry of the 1880s, however, never took off. Power engineers doubted that solar cells could generate large amounts of power. They were accustomed to working with heat to produce energy. No system could exist, they argued, which could generate usable amounts of energy without consuming significant quantities of material substances. Anything to the contrary, the engineering and scientific establishment maintained, violated the principle of the conservation of energy. Secondly, they felt that solar energy could not produce enough heat to generate large amounts of energy. The measurement of the solar constant proved this.

Contemporary solar cell promoters could not raise theoretical arguments against these objections. The reason why solar cells work—the photovoltaic effect—lay beyond the theoretical framework of classical physics. Such knowledge was far beyond anyone's reach at the time. They could only look in awe, as did Werner Siemens, the renowned German industrialist, inventor and scientist, pronouncing Fritts' work as "of the greatest scientific significance since there is here presented to us, for the first time, the direct conversion of the energy of light into electrical energy."

Only after the general acceptance of quantum mechanics—which explained scientifically how solar cells produce electricity directly from sunlight—did serious interest in solar cells recommence among engineers and scientists. By the early 1930s scientists had rediscovered the selenium solar cell, renewing Fritts' dream of producing fuelless electricity for commercial purposes. A 1931 article in *Popular Science* suggested that once these cells were refined they might power "a huge solar electric station at a cost no greater than would be required to build a hydro-electric station of the same capacity." In a similar vein, *Outlook and Independent* prophesized that these cells were "within range of practical large-scale power generation." Unfortunately, these new selenium cells were tied to the same low theoretical electrical output that Fritts and other early researchers faced. The 1930s also saw a revival of the interest in cuprous oxide/copper solar cells, but the cells were inefficient because of the location of the active junction.

Through the early 1950s selenium cells remained the most effective and reliable of the solar cells. The best cells could transform 1% of all incoming sunlight into electricity—hardly enough to be used as a power source. At this time, researchers at Bell Telephone Laboratories were seeking a dependable alternative energy source to power communication systems in isolated areas. Darryl Chapin, the leader of this research team, concluded that a solar-powered device would be the ideal solution and tried to develop a more efficient selenium cell, but to no avail.

Meanwhile, another Bell scientist, Cal Fuller, had been exploring ways of making silicon, a material similar to selenium, into a more efficient rectifier—a device which permits electrical current to travel in one direction. Fuller increased efficiencies by adding impurities to the silicon; when an outside voltage was applied, a stronger current flow resulted. The director of the rectifier program, Gordon Pearson, fortuitously exposed one of Fuller's improved rectifiers to light. To his surprise, Pearson recalled, "I noticed that it was very light-sensitive." A considerable amount of electrical current was generated.

Pearson recognized the obstacles Chapin faced and brought his discovery to his friend's attention. Soon Fuller and Chapin busied themselves refining the new solar cell. Pearson reminisced, "Although at the start we weren't going after a solar cell at all, upon this discovery, we turned it into a solar cell project." Chapin and Fuller found that this "rectifier" converted 4% of all incoming sunlight into electricity. Thus, their solar cell worked four times better than the best selenium cells.

Not content with the 4% conversion factor, they worked for several months trying to make the solar cell even more efficient. Pearson described the process as the "usual cut and try research." By May 1954 they announced the development of a solar cell with an efficiency of 6%. Pictures of Chapin, Fuller and Pearson with a solar-powered transistor radio were flashed across the nation and international borders. According to Chapin, their discovery "shook the whole world." *Business Week* reflected such enthusiasm, predicting that these solar converters would soon power phonographs, fans and lawnmowers. The magazine even talked of a solar convertible car that would use "a solar cell to guide it down a white line by remote control." Since there would be no need for steering, the article concluded, "all the riders could sit comfortably in the back seat and, perhaps, watch solar-powered TV."

Even though Fuller and Chapin began building cells with efficiencies as high as 15%, they took a more sober view. "We tried to avoid making too much claim for it," Chapin explained, "because we knew it was in the laboratory stage, it was an expensive process, and there was a lot to be done before we could speak of lots of power." Unfortunately, high-purity silicon, costing $80 per pound, was required for efficient cells. Furthermore, a good cell had to be handcrafted. Each cell had to be sliced manually.

Many envisioned acres of solar cells supplying the world with cheap, clean energy. However, the cost of these cells kept this vision in the realm of science fiction. Instead, as Chapin pointed out, "Because we didn't want to claim too much and embarrass ourselves with how little we had, we emphasized small uses." The first practical application of these cells was to power a telephone repeater system in rural Georgia. The system, combined with battery storage for nighttime supply, worked without problems. However, after an economic analysis, it was shown not to be competitive with conventional sources of energy.

Just as these cells were about to be consigned to the curiosity heap, the space race resuscitated their utility. Satellites and other space vehicles required a long-term power source, and power lines, of course, could not be called into service. The only alternatives to the solar cell were cumbersome fuel systems or batteries. Solar cells in space require no storage because of the continual sunlight. In this special niche, solar cells proved cost-effective—and they were the lightest per watt source of energy. The American space program created the solar cell industry. Virtually all the satellites, from Vanguard to Skylab, were powered by solar energy.

Terrestrial applications of solar cells remained untried and unexplored. As early as the mid-50s, the *New York Times* suggested that "It may well be . . . we ought to transfer some of our interest in atomic power to solar power." Why government never funded research of cheaper, more efficient cells mystified some. Perhaps the complacency of a nation hypnotized by the spell of limitless, cheap fuel made government ignore this potentially valuable energy source? Or was the cause of its disinterest an almost religious belief in nuclear as the energy savior of the future?

Today, fuel is no longer cheap, and we know that nuclear energy will not save us. Electricity from solar cells is a step toward energy independence. Richard Komp's practical guide to photovoltaics tells us how we can effectively put the sun's power to use in generating electricity.

John Perlin
Santa Barbara, California
Co-Author, *The Golden Thread:
2500 Years of Solar Architecture & Technology*

PREFACE

Since the first edition of *Practical Photovoltaics* was published, both the alternative energy field and the photovoltaics industry have experienced considerable change. While the pressing need for energy conservation and alternative energies continues, the panic has abated and government involvement in research and development has declined drastically. Nonetheless, the photovoltaics industry has become international in scope with an existence that goes far beyond the programs set up by the Department of Energy or other government agencies. More solar cell companies, however, have been bought out by oil companies, which now essentially control the photovoltaics industry.

The amorphous silicon solar cell has come into its own, powering millions of calculators and other small devices; polycrystalline silicon blocks are being made into solar cells in Germany and Japan, as well as in the United States. Sales of solar cell modules and systems are growing fast, increasing from about five megawatts per year when the first edition was published to better than 20 megawatts a year currently.

Solar cells are now an important product, the industry is maturing, and the future looks bright. Photovoltaic cells are becoming more efficient as their cost is coming down. Research is continuing, not only to better understand the physics and chemistry of the present crystalline and amorphous solar cells, but also to develop entirely new cell concepts, such as high-efficiency tandem cells.

To reflect the changes of the past few years, a chapter examining "The Photovoltaics Industry" has been added to this revised edition, and "The Future of Photovoltaics" has been updated and expanded. The chapters on building and installing systems report recent developments and improvements. Throughout the book, data has been brought up to date and new information added. *Practical Photovoltaics*, in its second edition, is now more comprehensive, useful and current, and I look forward to its being as well received as the first.

—RJK

PREFACE
To the First Edition

Energy independence has become a national goal. It is imperative that we gain direct control over our energy sources, rather than rely on dwindling fuel supplies and uncertain political alliances. The only truly independent sources of energy are the constantly renewed sources surrounding us: the wind, flowing water, growing plants, and the sun—the ultimate energy source. More and more, we are turning to the renewable resources for our many energy needs.

Today it is relatively common to heat our homes and water with the sun. We have also developed solar techniques to dry crops, make steam for industrial processes, and even cool our buildings. But the most desirable use of the sun's energy, and the one most important to our future, is the production of electricity. Photovoltaic cells, or solar cells, can capture the sun and convert it into electricity. They are rugged, reliable, long-lasting, have no moving parts, and give off no chemical by-products. Furthermore, they can be used almost anywhere, with no damage to the environment.

Many people are interested in photovoltaic cells, but are deterred by the high price and by their lack of knowledge on how to use them. Some have the impression that solar cells are an exotic source of power best left to the space program or the distant future. This is not so. Photovoltaic cells are practical right now for many purposes, and it is quite simple to learn how to use them. Anywhere electricity is needed and not available from a central utility, solar cells may be the best way to furnish that electricity.

The idea for this book grew out of my series of solar cell workshops, the first of which were held at Skyheat, a small nonprofit research foundation in Southern Indiana. Workshop participants actually assemble their own solar arrays from surplus solar cells and commonly available construction materials. It became apparent that a manual would be very useful for these ongoing workshops. This book is the result.

Practical Photovoltaics is more than a workshop manual or a do-it-yourself book. It serves both the technical and the nontechnical reader. The first section is devoted to a general description of solar cells and modules. Then a technical section details how to use solar cells and offers step-by-step instructions for assembling solar arrays. Chapters on how solar cells work and how they are made follow. Finally, a section on the photovoltaic industry, new scientific developments and future prospects gives an overview of this fascinating application of solar energy.

—RJK

Richard J. Komp received his Ph.D. in physical chemistry from Wayne State University. He began his research in photovoltaics as a physicist with Xerox Corporation and has since taught physics and conducted research at Western Kentucky University and Wayne State University. Research interests have included the mechanism  of dye photosensitization of selenium and zinc oxide; imaging systems; organic dye semiconductors; and cuprous oxide solar cells.

Dr. Komp designed and built Skyheat, a nonprofit solar research complex in Southern Indiana. He has recently constructed a completely self-sufficient home, research and manufacturing facility on the coast of Maine.

In addition to research and consulting, the author conducts hands-on workshops nationwide, including a series on photovoltaic cells. Dr. Komp is Vice-President, Research and Development, of SunWatt Corporation.

ACKNOWLEDGMENTS

I gratefully acknowledge the aid of Dr. Dan Trivich, Dr. Edward Wang, and others who furnished a good deal of the information on photovoltaic cells and inspired me to develop this book.

I also am happy to thank Vicky Patton, who typed the manuscript, Lawrence Komp, who drew the diagrams, David Ross Stevens, who furnished many of the workshop photographs, and the manufacturers who kindly supplied illustrations.

Finally, I wish to acknowledge the help of my workshop participants, who have assisted in the development and testing of procedures used to make practical use of photovoltaics.

CONTENTS

Chapter 1
Solar Cells and Arrays

SOLAR CELLS

Solar cells are solid-state devices that absorb light and convert light energy directly into electricity. They do this completely inside their solid structure and have no moving parts.

When you look at a solar cell, you see a metallic-blue or black disc covered with thin silvery lines. Figure 1.1 shows typical silicon solar cells and the silicon materials from which they are made. The area that does the actual work of converting sunlight into electricity is the dark metallic area, since dark colors absorb light more readily. The silvery lines are **front contact fingers** and are used to make the electrical contact to the front of the cell. The fingers are fine so as to block out as little sunlight as possible. The back contact is a solid metal layer designed to reflect light back up through the cell as well as to make a good electrical contact. For an explanation of how solar cells work, see Chapter 5.

Electrical Characteristics

In order to use photovoltaic cells we should have a basic understanding of their electrical characteristics. The cells, when illuminated, act somewhat like a battery in that they produce a voltage

1

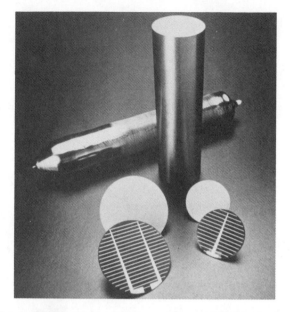

FIGURE 1.1–Silicon ingot, cylinder, wafers and finished solar cells.
(Courtesy Jet Propulsion Laboratory, California Institute of Technology)

between front and back. This voltage is developed across a current from this cell, just like from a battery, but the amount of current is limited by the amount of light falling on the cell.

We can use a simple circuit (Figure 1.2) to explain the electrical behavior of solar cells in more detail. A load resistor connects the front and back of a solar cell. The resistance of the load can be varied from a short-circuit zero resistance to a very high valence. Two meters—a voltmeter and an ammeter—measure the voltage developed across the cell and the current passing through the load.

If sunlight shines on the cell when the load resistance is set very high (or the load disconnected), giving essentially infinite resistance, the voltmeter will read a maximum voltage. For example, with the typical silicon solar cell, the voltmeter would read 0.58 volts. This is called the **open-circuit voltage**. No current is being drawn from the cell under these conditions. Conversely, if the load resistance is made zero, we will short-circuit the cell (which does a solar cell no harm whatsoever) and draw the maximum possible current from the cell. This current is directly proportional to the amount of light falling on the cell and is called the **short-circuit**

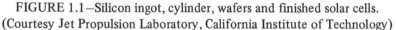

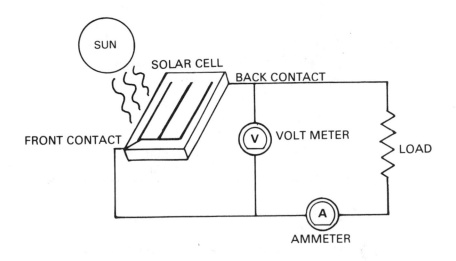

FIGURE 1.2—A simple circuit to test a solar cell.

current. It is possible to adjust the load resistor between these two extremes and measure the voltages and corresponding current produced by the cell under different load conditions. Figure 1.3, a **current-voltage curve** or **I-V curve**, is a plot of the results of such an experiment. This figure plots the current on the vertical axis versus the voltage on the horizontal axis. The short-circuit current is shown on the current axis at zero voltage. As the load resistance increases, causing the voltage output of the cell to increase, the current remains relatively constant until the "knee" of the curve is reached. Then the current drops off quickly, with only a small increase in voltage, until the open circuit condition is reached. At this point, the open-circuit voltage is obtained and no current is drawn from the device.

The power output of any electrical device, including a solar cell, is the output voltage times the output current under the same conditions. The open-circuit voltage is a point of no power, since the current is zero. Similarly, the short-circuit condition produces no power since the voltage is zero. The maximum power point is the best combination of voltage and current and is shown in Figure 1.3. This is the point at which the load resistance matches the solar cell internal resistance.

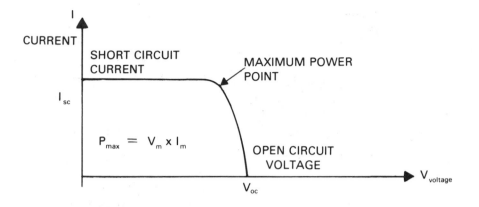

FIGURE 1.3–Current-voltage (I–V) characteristics of a
silicon photovoltaic cell.

Figure 1.4 shows a series of I-V curves for a solar cell under
different amounts of sunlight. The peak power current changes
proportionally to the amount of sunlight, but the voltage drops
only slightly with large changes in the light intensity. Thus, a solar
cell system can be designed to extract enough usable power to
trickle-charge a storage battery even on a cloudy day.

Cell Performance Ratings

To compare the performance of different solar cells, the cells
are rated at specified amounts and types of sunlight. The most
common rating parameter for terrestrial cells is called **Air Mass 1
(AM1)**. This is the amount of sunlight that falls on the surface of
the earth at sea level when the sun is shining straight down through
a dry, clean atmosphere. The Sahara Desert, at high noon, with
the sun directly overhead, is a close approximation of this condition. The sunlight intensity under these conditions is very close to
1 kilowatt per square meter (1 kW/m^2). A closer approximation to
the sunlight conditions usually encountered is **Air Mass 2 (AM2)**,
an illumination of 800 W/m^2.

Table 1.1 gives sizes and typical output currents for a number
of silicon solar cells produced by various manufacturers.

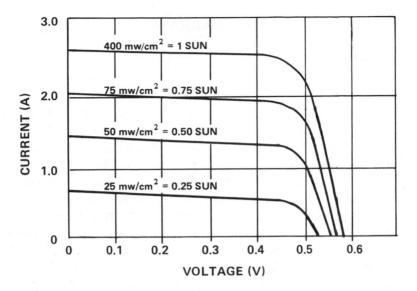

FIGURE 1.4–Series of I–V curves showing solar cell characteristics versus sunlight intensity.

Table 1.1

Sizes and Rated Outputs for Representative Silicon Solar Cells

Cell Size	Output Current (amperes)	Manufacturer*
3-inch round (75 mm)	1.3	Solec
one-half of a 3-inch round	0.6	Applied Solar Energy
2 x 6 centimeter, rectangular	0.4	Applied Solar Energy
4-inch round (100 mm)	2.1	ARCO Solar
4-inch square	2.1	Semix
one-fourth of a 4-inch round	0.53	Solenergy
2¼-inch round	0.6	Solenergy
2-inch round	0.5	Photowatt
5-inch round	2.9	Photowatt

*While some of these companies are no longer active, the cells are still available on the surplus market.

SOLAR ARRAYS

Series and Parallel Strings

The maximum useful voltage generated by a solar cell is on the order of 0.5 volt. This is not a very high potential, and not many devices can operate on such a small voltage. However, solar cells can be connected in **series** (the back contact of one cell connected to the front contact of the next) to obtain a higher voltage. Figure 1.5a shows such a series string, which is called a **solar array** (or **solar module**). (The terms, solar array and solar module, are sometimes used interchangeably, but usually a solar array is a set of one or more modules.)   A typical solar module will have 32 cells in series, producing an open-circuit voltage in bright sunlight of around 18 volts, or 16 volts when producing its maximum power. Such a module is designed to charge a 12-volt storage battery. It is important to remember that the total current output of a series string of cells is the same as that of a single cell.

To increase the current output, solar cells are wired in **parallel** (the back contact of one cell connected to the back contact of the next, with the front contacts similarly connected). In a parallel system, such as shown in Figure 1.5b, the total current is the sum of the individual current outputs of the cells, but the total voltage is the same as the voltage of a single cell. Usually, individual cells are wired in series, forming an array, to obtain the desired output voltage, and arrays are wired in parallel to obtain the desired current. Chapter 2 details this procedure.

Solar cell manufacturers produce a variety of standard modules that can be purchased and combined into an array to meet just about any conceivable voltage and current needs (see Figure 1.6). Appendix A lists manufacturers and suppliers.

Array Construction Techniques

Since silicon solar cells and arrays were first developed for the space program, the construction techniques were worked out with an emphasis on reliability. In most cells the front contact finger

BACK TO FRONT

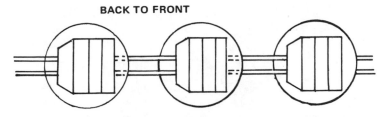

(a) SERIES CELL ARRANGEMENT

BACK TO BACK

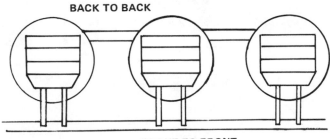

FRONT TO FRONT

(b) PARALLEL CELL ARRANGEMENT

FIGURE 1.5—(a) Series string of solar cells increases voltage.
(b) Parallel string of solar cells increases current.

patterns are designed with some redundancy so that a break in the pattern or a bad solder joint doesn't stop the entire cell from working. Figure 1.7, a Motorola square cell (100 mm on a side), is a good illustration of this design technique. The cells are fastened into series strings by sweat-soldering thin copper ribbons over the three main crossbars. Then the ribbons are sweat-soldered onto the back of the next cell. Even if a cell cracks, the two pieces are still fastened in parallel and the electrical performance is virtually unchanged.

FIGURE 1.6—State-of-the-art solar cell modules for
intermediate load center applications.
(Courtesy Jet Propulsion Laboratory, California Institute of Technology)

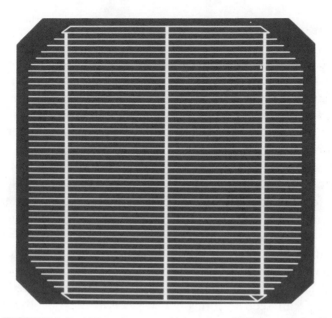

FIGURE 1.7–100-mm-square (4-in.-square) silicon solar cell.
(Courtesy Motorola Semiconductor Group, Phoenix, Arizona)

The Motorola cell pictured here has nickel fingers which are tinned by an industrial wave-soldering machine. Other cells are made with silver-plated and palladium-silver fingers. Newer techniques using silkscreen printing to deposit the fingers are being used now. (Finger contact design is a very important area of research in developing new lower-cost cells.) Since the fingers do block light from the active cell material, they must be as thin as possible, yet able to withstand mechanical strain while providing a good electrical contact to both the cell and the interconnecting strips. Corrosion and mechanical failure of the finger contacts are the most common causes of cell failure, so noncorrosive metals must be used. However, the metals must have the proper electronic behavior in contact with silicon. Most cells are soldered together to make arrays, but some manufacturers are now trying to spot-weld the interconnections. Many solar modules come equipped with a built-in **blocking diode**. This diode protects a storage battery against drainage by preventing the current from running backwards through the solar cell array. Blocking diodes will be discussed further in the next chapter.

Encapsulation

Cells must be sealed to protect them from the environment and to support them in the array. The two most common encapsulants are transparent silicone rubber and a butyral plastic similar to the plastic layer used in automobile safety glass. A top cover of plastic or tempered glass is normally added to the array to offer better protection against the elements. Glass covers are more scratch-resistant and remain transparent longer, but they do not flex as much as plastic covers. This flexibility makes plastic modules more suitable for use on sailboats and camping vehicles. The plastics used in the covers are acrylic and polycarbonate. Of the two, polycarbonate is the more damage-resistant and can withstand higher temperatures, but acrylic is more flexible and less expensive.

To bring the electrical wires out of the panel requires some care in design. The electrical connectors must be very well sealed; otherwise, water can seep in along the wires and corrode the interior contacts. Also, the connectors have to be very secure so the wires cannot be pulled out or put strain on the solar cells. Usually the connectors are a pair of contacts that protrude from the back of the panel and are sometimes encased in a small junction box. Some modules, however, are designed so that the wires project from the side, which allows the panel to rest flat on a support surface (for example, a roof or sailboat cabin top).

Array Failure Mechanisms

Joints Between Cells

The silicon solar cell is a remarkably rugged device. The cell itself will withstand a great deal of abuse and extreme temperatures, both high and low. It is also impervious to most corrosive chemicals. When a solar array fails, the problem usually lies with something other than the cells. For example, the first solar array I constructed failed prematurely after four years of operation on the deck of a small sailboat. In this case the main problem was corrosion from salt water that had leaked into the case. Some of the joints between cells had also failed mechanically. The cells had been fastened in series by a technique called **shingling**. The front edge of one cell is overlapped by the back edge of the next, similar to roof shingles, and the edges are soldered directly together. No copper jumpers are used to interconnect the cells except from one row to the next. Since the entire row is one rigidly fastened unit, any flexing or thermal expansion puts a great strain on the adjoining edges, pulling the top contact loose from the underlying cell.

Producing a good mechanical bond between the silicon cell material and the nickel or silver top finger contact is difficult, and it is surprisingly easy to pull the fingers right off the cell surface by jerking on the jumper strip soldered to the fingers. The back contact covers the entire back of the cell and, while it isn't actually any stronger, the greater surface area helps make the cell more durable.

Too high a temperature during the soldering of contacts to the cells will also loosen the fingers from the silicon, while too low a temperature will cause a "cold" solder joint that will quickly fail. The exact composition of the solder and the flux, and the cleanliness of the operation greatly influence the reliability of the joints.

Encapsulant Problems

If the thermal expansion of a rigid encapsulant differs sufficiently from that of the cells, the encapsulant can pull loose or the cells can even crack. This has been a problem with cells embedded in polyester resins.

Direct sunlight creates a very harsh environment and can cause chemical reactions in encapsulants and cover materials, turning them cloudy or yellow. While this may have no effect on the cell itself, it will cut down on the total light reaching the cells and lower the system efficiency. These reactions are accelerated by the high temperatures arrays can attain under full sun in the summer. The cells actually can withstand much higher temperatures than any other part of the array, although cell efficiency drops when the temperature rises. It is best to keep the cell temperature below 60°C.

Shunt Diodes

A cell can be permanently damaged if a large reverse voltage is applied to the electrodes. A solar cell is an electrical rectifier: it passes current in only one direction. In the dark, the silicon solar cell acts just like any other silicon diode rectifier. However, in the manufacture of solar cells, no real attempt is made to build in an ability to withstand a large reverse bias, since normally the cell is not used in this way. But there is one circumstance under which this reverse voltage condition can occur: when one cell is shaded while the rest of the cells in a series string are in sunlight. The current though the string immediately stops, and the sum of all the open-circuit voltages of all the other cells shows up across the shaded cell. If the cell cannot withstand this voltage, it will break down electrically and begin to conduct. The resistance heating effect of the current can make a cell hot enough to melt the solder connections and destroy the fingers.

Sooner or later, such shading will occur; for example, a leaf falls on one cell, or a tree casts a shadow on an array. Since it is impossible to prevent such occurrences, it is necessary to take precautions so that shading will not destroy a cell. With 32-cell arrays, reverse bias breakdown is rarely a problem, as virtually all cells will take 20 volts or more in reverse bias. But if higher voltage strings are needed, shunt diodes should be installed. Chapter 2 describes how to wire shunt diodes, as well as other practical details of connecting solar arrays. Some manufacturers have constructed experimental cells with the shunt diode built into the cell structure. With these cells, no external shunt diode is necessary.

In spite of the potential troubles described above, solar cell arrays, if properly designed and installed, can last a remarkably long time. The first solar arrays made by Bell Laboratories and originally installed in Americus, Georgia, are still operational after decades. A minimum of 25 years of reliable operation can be expected from a solar electric installation.

Concentrator Arrays

One way to use expensive solar cells more efficiently is to concentrate more light onto the cells. A detailed description of the numerous schemes that have been devised to accomplish this is beyond the scope of this work. Rather, we will discuss the general principles and give examples of each type.

The first parameter that can be used to characterize a concentrator system is the **concentration ratio**. This is simply the ratio between the area of the clear aperture, or opening through which the sunlight enters, and the area of the illuminated cell. Figure 1.8 shows a simple lens-type concentrator similar to a magnifying glass used to start fires. The concentration ratio is the area of the lens divided by the area of the cell. If the lens is 200 mm in diameter, its area is 314 cm^2; if the cell is 50 mm in diameter, its area is 20 cm^2 (for a circle the area $A = \pi r^2$). The concentration ratio is $314/20 = 16$ to 1 (sometimes expressed as a concentration ratio of 16 suns). This means that the cell receives 16 times the light (minus the reflection and absorption losses in the lens, which can be very small for a coated lens) and should put out 16 times the power of a similar cell used without the lens. The lens does not have to be of fine optical quality and could be of molded plastic,

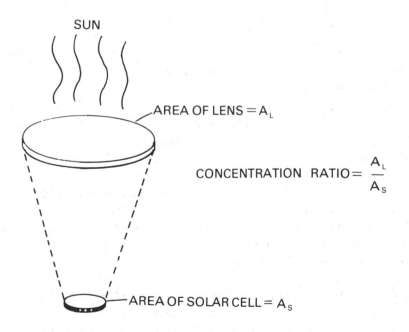

FIGURE 1.8—Diagram of a simple lens-type concentrator.

so that such a system would appear to be much cheaper than the 16 cells it replaces. The reality is not quite so simple, however.

First, with all the extra sunlight pouring into it, the cell is going to heat up. With a concentration ratio of 3 suns or more, some way must be arranged to cool the cell. Sometimes, simple cooling fins mounted to the back of the cell will suffice, but in hot climates or with higher concentration ratios, water or forced-air cooling becomes necessary. The excess heat removed from the cells can be saved and utilized. A system that produces usable heat as well as electricity is called a **hybrid system**. Hybrid systems can be very cost-effective in situations where neither the heat nor the electricity generated separately justifies a solar system. Hybrid systems are discussed later in this chapter and again in Chapter 4, which contains step-by-step instructions for hybrid array construction.

Second, the currents produced by the photovoltaic cells in concentrator systems can become quite large. (Currents in excess of 100 amps have been produced by a single cell.) In order to handle these massive currents, special cells have to be constructed. Usually these cells are thinner than the normal cell and have a much smaller cell resistance. The finger pattern and top contact for the cell must be carefully designed to carry such currents without blocking too much of the cell's surface area. Figure 1.9 shows a concentrator cell capable of handling a concentration ratio of up to 200 suns. Naturally, such cells are more expensive than the usual solar cell, but each cell produces so much power that the extra cost is justified. Ordinary solar cells can be used up to a concentration ratio of 3 or 4 suns, provided they can be cooled adequately.

Third, since a concentrator system must point directly at the sun in order to work, a system must be devised to pivot and follow the sun as it travels daily from east to west. The pivot must also raise and lower the angle to allow for the seasonal change in the sun's path in the sky—high in the summer and low in the winter. The simple concentrator shown in Figure 1.8 now has a complex tracking mount. The mount could be moved by hand, but that would require constant adjustment, or it could have a clockwork mechanism to track the sun automatically. The most sophisticated systems have photocell detectors and feedback mechanisms to constantly correct the system and keep it pointed properly.

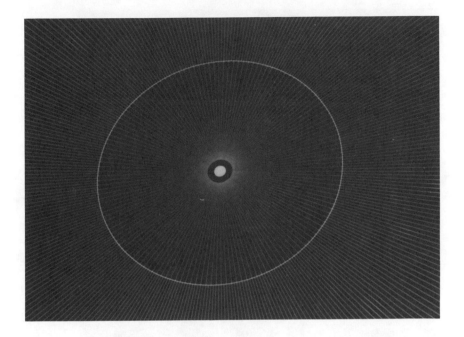

FIGURE 1.9–Concentrator cell capable of handling ratios of 200 suns. (Courtesy Applied Solar Energy Corporation, City of Industry, California)

2-Axis Tracking

A mount that is capable of pivoting both daily and seasonally to follow the sun's path is called a **2-axis tracking system**. A large number of 2-axis tracking systems have been built experimentally and several are available commercially. Some units use a small concentration ratio of 3 to 1, so a simple cone reflector is sufficient and large tracking errors can be tolerated. If the unit also has forced-air cooling of the cells, the hot air produced can be put to some useful purpose.

Figure 1.10 shows another 2-axis tracking system that uses **fresnel lenses** as the concentrator elements. Fresnel lenses work as well for light concentration as the thick lenses they replace, but use only a fraction of the material and can be molded out of plastic. It is also possible to have a greater optical correction in the

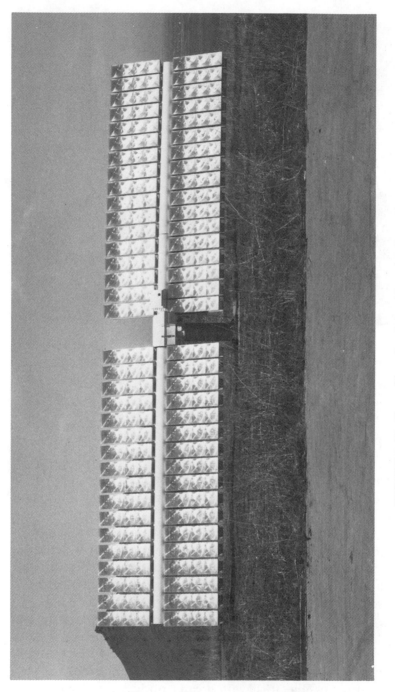

FIGURE 1.10—Fresnel lens 2-axis tracking system array. (Courtesy Martin Marietta Aerospace, Denver, Colorado)

outer edge of a fresnel lens so the distance from the lens to the cell can be shorter for the same concentration ratio. This particular fresnel lens concentrator system, produced by Martin Marietta Corporation, has been installed in a village in Saudi Arabia and has a 350-kW capacity, enough to supply the power needed by two villages and replace a set of diesel generators. The system, with a concentration ratio of 33 suns, has a particularly sophisticated tracking controller using coarse and fine sun sensors and a small computer. No provision is being made to collect the waste heat from the system, which is passively cooled.

1-Axis Tracking

The purpose of a solar concentrator is not to give a clear image of the sun, but rather to collect the sunlight that falls over a large area and project it onto the small area of the collector or array. The quality of the image doesn't matter as long as most of the sunlight going into the device falls onto the cells. Figure 1.11 shows a trough-type parabolic collector that does this in one dimension. This particular embodiment of the trough collector is called a **Winston Concentrator** and was developed by Roland Winston, at Argonne National Laboratories. Its great advantage— for smaller concentration ratios—is that it doesn't have to track the sun daily for the seasonal variation in the sun's position. Table 1.2 gives the **acceptance angle** and frequency of adjustment for different concentration ratios. The acceptance angle is the total range of sun positions over which the sunlight will still be collected by the mirror system and focused onto the solar cells. This table assumes that the axis of the trough collector is horizontal and pointing east and west.

Such a system, while simple, does not effectively collect the sunlight early in the morning or just before sundown, as at these times the sun is shining more or less lengthwise down the collector. Another version of the linear concentrator is oriented north and south and tilted to the south at an angle equal to the latitude of the installation. The system tracks the sun during the course of a day but needs no adjustment for seasons, so the mount and tracking mechanism can be quite simple. Also, connecting the cooling

FIGURE 1.11–Trough-type parabolic collector, similar to that developed by Winston, mounted on the roof of the Raymond Frady residence near Mifflin, Indiana. This type of hybrid concentrator photovoltaic system is particularly suitable for remote residential applications.

Table 1.2

Acceptance Angle and Adjustments Needed Per Year for Different Concentration Ratios for a Winston Concentrator

Concentration Ratio	Acceptance Angle (degrees)	No. of Adjustments Per Year	Shortest Period Between Adjustments (days)
2	30	0	–
3	20	2	180
4	14.5	4	35
6	9.6	10	26
10	5.8	82	1

fluid pipes or air ducts to the system is greatly simplified if a hybrid or **total energy system** is desired.

Whatever solar cell configuration is used, simple flat plate or concentrator, hybrid or electricity-collecting only, the inherent simplicity of operation of a solar cell makes the system easy to install and maintain. The only connections to the solar cell array are a simple pair of wires, giving the user a great deal of design flexibility. The next chapter explains exactly how solar cells are used.

RECOMMENDED READINGS

Backus, Charles, E., Editor. *Solar Cells* (New York: Institute of Electrical and Electronics Engineers, 1976).

Commoner, Barry. "The Solar Transition," *The New Yorker* 55: 53–54, Part I (April 23, 1970); 55:46–48, Part II (April 30, 1979).

Fan, John. "Solar Cells: Plugging into the Sun," *MIT Technology Review* 80:14–15 (August 1980).

Institute of Electrical and Electronics Engineers. *Spectrum* 17: (February 1980). A special issue devoted to photovoltaics.

Merrigan, Joseph A. *Sunlight into Electricity* (Cambridge, MA: The MIT Press, 1975).

Pulfrey, David, Editor. *Photovoltaic Power Generation* (New York: Van Nostrand Reinhold Co., 1978).

Roberton, Ed. *The Solarex Guide to Solar Electricity* (Rockville, MD: Solarex Corporation, 1979).

Winston, R. "Principles of Solar Concentrators of a Novel Design," *Solar Energy* 16(2):89 (October 1974).

Chapter 2
Using Solar Arrays

Solar cells are not difficult to use. In this chapter, we will outline the steps of deciding what kind of solar cell system is needed and then show the practical details of installation and operation.

When are solar cells worth using? Before you can answer that question, you have to know two things: one, how much it will cost to do the job with solar cells and, two, what the alternatives would cost. Solar cells are expensive. Just how expensive we will see in a moment, but at their current price, they cannot possibly compete with electricity delivered by a power company. So if regular utility electricity is available, the economics do not justify using solar cells for most applications. However, if you cannot simply plug into a power company for a certain application, then solar cells may be the cheapest way to generate the needed electricity. Solar cells are proving to be the best source of electrical power in remote locations where regular power is unavailable; on moving equipment, campers or boats; and for small temporary power needs.

The photographs on the following pages, Figures 2.1 through 2.10, show some of the many ways solar arrays have been used. Some of these uses, like the radio station in Bryan, Ohio, are strictly for demonstration purposes and to gain experience in generating large amounts of solar power. But in many of the examples, solar systems are doing the job better than any alternative could.

FIGURE 2.1–Photovoltaic-powered national park outhouse.
(Courtesy Motorola Semiconductor Group, Phoenix, Arizona)

FIGURE 2.2–Solar cell battery
charger installed on a cruising
sailboat.   (Skyheat photo)

FIGURE 2.3–**Solar Challenger**, designed and built by a team headed by Dr.
Paul Macready, is capable of 100-mile
flights, using solar energy as its sole
power source. Over 15,000 solar cells
are wired into its wings and tail. (Courtesy E. I. du Pont de Nemours and
Company, Inc., Wilmington, Delaware)

FIGURE 2.4–A stand-alone photovoltaic system that operates a pump to supply water to cattle.   (Courtesy New Mexico Solar Energy Institute, Las Cruces, New Mexico)

FIGURE 2.5–Photovoltaic-powered radio station, WBNO, near Bryan, Ohio. (Courtesy Jet Propulsion Laboratory, California Institute of Technology)

FIGURE 2.6—Small, inexpensive photovoltaic-powered battery chargers are now available. This unit recharges flashlight batteries; larger units charge 9- and 12-volt batteries. (Courtesy SunWatt Corporation)

FIGURE 2.7 — 28-kW array for irrigation and crop drying located near Mead, Nebraska. (Courtesy Jet Propulsion Laboratory, California Institute of Technology)

FIGURE 2.8–Photovoltaic-powered forest lookout station. (Courtesy Jet Propulsion Laboratory, California Institute of Technology)

FIGURE 2.9–The Sunrunner, a PV-powered vehicle designed and built by Greg Johanson and Joel Davidson, has attained peak speeds of 24 mph on test runs, with speeds of over 40 mph possible.   (Courtesy Greg Johanson)

FIGURE 2.10–7.3 kW Solarex array powers this residence in Carlisle, Massachusetts.　　　　　　　　(Courtesy Solarex Corporation, Rockville, Maryland)

CALCULATING LOAD

The first step in determining what kind of solar array to use is
to decide what you want it to do. You must figure out what the
load on the solar cell array will be. First ascertain what voltage
your load needs. This task is easy for portable or marine systems,
since these systems operate from a **DC** supply with storage batteries. The main objective will be to keep the storage batteries
charged. A marine system will most likely be 12 volts, although
some larger boats have 24- and even 120-volt AC power. For the
present example, let's assume a 12-volt system. As mentioned
earlier, several manufacturers make a 32- to 36-cell module that
is designed to charge a 12-volt battery system. These systems come
in a wide range of current outputs and physical sizes.

The next step is to calculate how much power the system will
need. This can be expressed in **ampere-hours (AH)**. A current of
one ampere running for one hour is one ampere-hour. You must
find out or estimate how much current is drawn by the device
you want to use and also estimate how long, on the average, each
device will run. For example, if a lamp draws 2 amps and you
want to use it 3 hours per night, that's a total of 2 x 3 = 6 AH of

electricity used per day. You can thus calculate the total load for the various devices you intend to use. If the solar cell array is to power only one or two devices, this is a fairly simple procedure, but for a summer cottage or sailboat, such calculations may prove complicated, and it may be impossible to predict exact usage.

A shortcut can be used if you are installing the solar array on an existing 12-volt system that presently runs off storage batteries. Suppose you notice that the battery system runs for about 3 days without any charging before the voltage gets unacceptably low. This means that you have probably discharged the batteries to 3/4 of their rated capacity. Ascertain the ampere-hour capacity of the battery and take 3/4 of the capacity as the ampere-hours used in the time period. Suppose the battery in this example is rated at 100 AH. You used 3/4 x 100 or 75 AH of power in 3 days, which comes to 25 AH per day.

At some point, you must decide what type of conventional back-up to use, and what percentage of the total requirements the system will furnish. Operating a large electrical system 100% by the sun is a costly matter. The system would have to be oversized to compensate for a long cloudy spell or for the one time expected load is slightly larger than expected sunlight. The storage system would have to be large enough to take care of the last few percent of demand. In a small system, in which a reasonably sized battery system can operate the load for weeks, the vagaries of weather average out and the sun can be relied on to furnish all the power. Lighted navigation aids and small radio repeaters can operate this way. *The New Solar Electric Home* discusses this and other aspects of installation and use in much more detail.

SIZING THE SOLAR ARRAY

Once the average number of ampere-hours per day for the solar array has been determined, the next step is to choose an array large enough to furnish the average power needed. The average amount of sunlight available in different parts of the world varies widely and so must be taken into consideration along with the average amount of light available on different days of the year.

Researchers at Motorola have performed a computer analysis of these factors and have devised a simplified procedure that will give a good estimate of the system size needed for a particular application. The step-by-step Motorola method is included below (courtesy of Motorola Semiconductor Group, Phoenix, Arizona).

Sizing Methodology

Matching an electrical load to a photovoltaic array is an involved process which is greatly simplified with the aid of a computer program. The following procedure offers a good estimate of system size and is based on computer analysis.

Step 1

Determine load requirement. Calculate the daily ampere-hour requirements of the load by multiplying the load current in amperes by the number of hours per day used. In the case of a system which operates in more than one mode during the day, for example, transmit or standby, calculate the ampere hours per day required for each mode and add the resultant figures.

Example:

Transmit	10 amps for 2.5 hr/day = 25 AH/day
Receive	1.3 amp for 9.5 hr/day = 12 AH/day
Standby	0.5 amp for 12 hr/day = 6 AH/day
	Total 43 AH/day

Note: Array size and array cost are directly proportional to load!

Step 2

Select desired safety factor:

10% for an attended site
15% for an unattended, accessible site
20% for an unattended, inaccessible site
30% for an unattended, inaccessible site of a critical nature

The safety factor accounts for variables and losses in the system. Weather data is the greatest variable since solar irradiation varies from year to year and from site to site. Even locating the photovoltaic system 30 miles from the weather reporting station could significantly change the solar irradiation received. In addition, the accuracy of the recording equipment is typically no better than 5 to 10%. Dust will accumulate on the modules between rain showers and reduce output up to 8% (after each rain, full output is restored.) Other losses include self-discharge of batteries and bias current of voltage regulators.

Step 3

Multiply the daily ampere-hour load as defined in Step 1 by one plus the safety factor selected in Step 2.

Example (using 15% safety factor):

43 AH x 1.15 = 49.45 AH (use 50 AH as the system load)

Step 4

Find the installation site on the location code map. Use Figure 2.8 for the continental United States and Figure 2.9 for all other worldwide locations. Note the area code for the site.

Step 5

From the system output (AH/day) defined in Step 3 and the area code defined in Step 4, use the appropriate multiplier to determine array size in watts:

Area A   AH/day x 3.08 = array watts
Area B   AH/day x 3.77 = array watts
Area C   AH/day x 5.20 = array watts
Area D   AH/day x 6.90 = array watts
Area E   AH/day x 9.77 = array watts

Example (use 50 AH/day and Laramie, Wyoming—Area B):

array watts = 50 x 3.77 = 188 watts (peak)

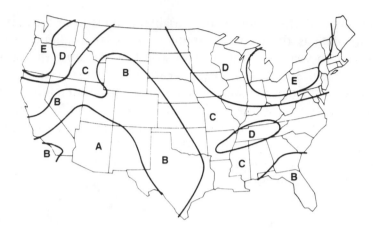

FIGURE 2.8–Location code map, continental United States

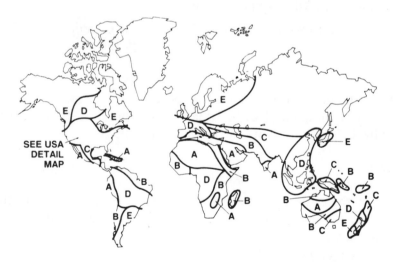

FIGURE 2.9–Location code map, worldwide.

These maps are used as an aid in determining the size of a solar array in different parts of the world. (Courtesy Motorola Semiconductor Group, Phoenix, Arizona)

Array output defined by this step assumes a 12-volt system and an array tilt angle equal to latitude.

Step 6

Calculate battery storage capacity by multiplying load requirements from Step 1 (not Step 2) by 9.6.

Example:

43 AH/day x 9.6 days = 413 AH

The number 9.6 is the result of 8 "no sun" days plus a factor of 1.2. Since battery capacity may degrade 20% in 5 years, this will ensure 8 "no sun" days capacity at the end of 5 years.

Step 7

Battery capacity must be adjusted for minimum temperature expected.

Example:

Assume a minimum temperature of 14°F and an available battery capacity of 78% of rated value (at 77°F). Since 413 AH (from Step 6) is needed at 14°C, then the rated battery capacity (at 77°F) would be:

Battery capacity = 413 AH/0.78 = 529 AH

Summary

From the above procedure, it has been determined that a 43-AH/day load will require a photovoltaic array of approximately 188 watts (peak) and battery storage capacity of approximately 529 AH. The total cost of such a system, including modules, batteries, voltage regulator, wiring and steel structure, is approximately $3500. When one considers that there are no operating costs, nearly zero maintenance, and no pollution, this system becomes very cost-effective.

While the system is relatively easy to use and will specify a solar array size that will certainly meet the needs of a small, steady load, the chosen array may be larger than justifiable if an auxiliary electric system is available or if it is possible to limit electrical power usage during long cloudy spells. In such a case you may decide to install one-half or two-thirds of the calculated array and see how satisfied you are with the results, planning to add more capacity as needed or as your budget allows. For a more detailed method of calculating the expected output of a solar array on a month-by-month basis in different parts of the country, see Rauschenbach's *Solar Cell Array Design Handbook.* **compuSOLAR** produces a computer program that performs these same calculations, available ready to run on several popular home computers.

The Motorola procedure calculates the peak wattage of array output needed. This is the output of the array to AM1 sunlight conditions. Some solar cell manufacturers express the output of their modules in terms of amperes instead of watts. If they are referring to the normal 32- to 36-cell module intended to charge a 12-volt battery, it is possible to divide the number of watts needed by 16 to get the peak array current needed. (Motorola specifies the wattage of their modules as power = current x 16 volts.) So in the example above, 188 peak watts/16 = 11.75 amperes peak current.

CURRENT COST OF SOLAR ELECTRICITY

To give an idea of how much solar electricity costs at the present time, here are sample calculations based on either: (A) the best purchase price by the federal government to date ($6 per peak watt); or (B) the best purchase price by a private individual for a single finished module (about $10 per watt).

For a person in Indianapolis, using the weather data from the Motorola computer program, one watt of solar cell capacity would produce about 1.2 kWh of electricity a year. Over the 20-year expected life of the array (the array might last much longer, but this is the standard expected lifetime), the system will produce 1.2 kWh x 20 years @ 24 kWh. For case A ($6/peak watt) the electricity costs $0.25 per kWh. For case B ($10/peak watt) the cost is $0.41 per kWh. This doesn't include any of the other system costs, such as those of support structures, wiring or storage batteries, which could easily double the final price per kWh. Even in Arizona, where one could expect to get as much as 50 kWh per year per watt of capacity, the cost of solar electricity cannot yet compete with conventional utility prices. But if solar cell prices drop and regular utility prices continue to climb, in a few years it might pay, especially in sunny climes, to install residential solar electric systems.

INSTALLING SOLAR CELL ARRAYS

Packaged solar modules are very easy to install. Most of these systems are designed to be installed outdoors without additional protection; all that is needed is a rigid support that won't sway in the wind or collapse under snow loads. Many manufacturers sell supporting frames designed to hold their modules, but it isn't difficult to design your own for a special application. Just remember, the array should last 20 years or more with minimum attention. Can the support structure do as well?

Large arrays can be mounted directly on a roof, or can even be part of the roof structure. Experimental solar roof shingles have been made and, if they become cheap enough, the extra cost of solar shingles above conventional ones would quickly be repaid in utility bill savings. Solar cells have been built into sailboat hatch covers and can be fiberglassed into the roof of a travel trailer.

Because of their ability to operate for years without any attention, solar cells can be permanently installed in all sorts of creative ways: in a car roof to trickle-charge the battery whenever the car is parked in the sun; as canopies or sun shades on golf carts to extend their range; on the entire top wing of a glider to run a small prop.

Array Orientation

In a permanent installation, it is important that the orientation of the solar cell array make the most effective use of the available sunlight. If the array is to be fixed in place, the most useful general orientation is facing due south and tilted at an angle from the vertical equal to the latitude. If the array is to be used primarily in the winter, or if that is the time when the short days and cloudy weather put the greatest strain on the system to meet the load requirements, the array can be tilted down, up to 15° further, to increase its efficiency. Presumably, the loss in efficiency in the summer is more than made up for by the increased amount of sunlight. If the system is used only in the summer, it could be tilted up 15° or so from the latitude angle. Figure 2.10 shows these three cases. The angle is not critical, however; a 15° change in the angle only changes the overall efficiency by about 5%. So if a roof's orientation is nearly correct, a couple of pressure-treated wood blocks might be all that is necessary to build the support.

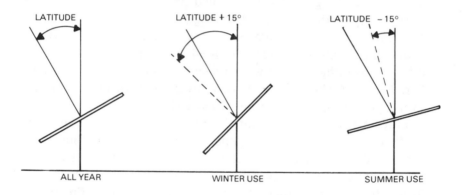

FIGURE 2.10—Array orientation showing suggested tilt angles for three different situations.

It is also possible, of course, to change the orientation of the array from time to time to make it operate more efficiently. A simple seasonal adjustment would be no more trouble than changing all your clocks to daylight savings time. Maybe in the future we will have national holidays when everybody ceremoniously readjusts their solar collectors for the season.

For concentrator arrays, some sort of adjustment is almost a necessity. Table 1.2 in Chapter 1 details the adjustments needed for different concentration ratios. The most precise adjustment would be a continuous tracking of the sun during the day. For a nonconcentrating array, this is probably not worth the mechanical complexity involved, but if the array is portable, someone could turn it occasionally to face the sun.

Wiring the Array

The wires running from the array to the storage battery (or the consuming device) should be carefully selected to withstand wind and weather. UV-resistant insulation and plated conductors are the best choice; the wires should be securely fastened or run in conduits so the wind can't whip them around and pull the connections loose from the solar array. If the array is fastened directly to a roof, it may be possible to make the connections to the back of the array and keep the wires completely out of the weather. The wire size should be sufficient for the peak current produced by the array. Appendix B is a table of current-carrying capacities for different wire sizes. A wire that is too big won't do any harm and will have more mechanical strength than the sizes given in the table for smaller currents.

Figure 2.11 shows a suggested wiring diagram for a 12-volt DC solar cell system. The lightning protection device would be necessary if the arrays are to be placed in an exposed location. The

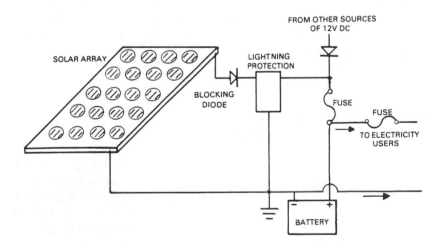

FIGURE 2.11–Suggested wiring diagram for a 12-volt DC solar cell stand-alone power system.

blocking diode is necessary to keep the battery from discharging through the solar array at night. With this blocking diode, it is not necessary to disconnect the solar cell arrays when some other device is being used to charge the batteries. A storage battery can deliver hundreds of amps into a short circuit, so a fuse should be placed as close to the battery terminal as possible to protect the rest of the wiring. This fuse should be big enough to carry the maximum expected current, but not oversized for the wire size used. The various devices that use 12-volt electricity can also be fused, of course.

COMBINATION SYSTEMS

A solar electric system can be combined with other alternative methods of generating electricity. A solar/wind combination is particularly good since, quite often, either one or the other is available. The wind system manufacturer can help you hook the storage batteries to the wind system. With a blocking diode in each system, they won't interact with each other, but will independently charge your battery whenever they produce a sufficiently high voltage. Figure 2.12 is a simple circuit that will light a small pilot light whenever the system is actually charging the battery. The circuit draws all its power from the charging system and uses no current from the battery.

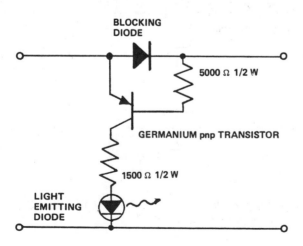

FIGURE 2.12—Simple circuit to indicate when battery is being charged. This circuit is powered by the solar array.

Solar/hydroelectric, solar/generator (gasoline-or alcohol-powered) or even solar/bicycle combinations are all possible systems. The Micro-Utility System installed at the Real Alternative Energy Fair held at Governor's State University demonstrated the simultaneous operation of a number of these systems.

INVERTERS

All the systems discussed so far are low-voltage DC. Most major appliances operate from 120-volt AC, however, and the electricity produced by solar cells must be converted to this form in order to be useful. A device that converts DC into AC is called an **inverter**. A great variety of these devices are commercially available. For the best of these, the quality of the current—in voltage, frequency and waveform—cannot be distinguished from that of utility power. Because of the wealth of information available on them already, inverters will not be covered in great depth here. The basic types and their advantages and limitations are given in Table 2.1.

Rotary Inverters

The rotary inverter is simply a DC motor turning an AC generator. These devices have reached a high state of development and the better ones will operate unattended for years. Older units are often available from surplus dealers for little more than the price of the scrap copper in them. Marine inverters are generally 60-cycle, but most aircraft inverters are 400-cycle and cannot be used with most appliances.

Solid-State Inverters

The cheaper solid-state inverters produce a square-wave output, and large-capacity commercial models are available. For many applications, *e.g.*, larger appliance motors, the square waveform is not detrimental; some work with capacitors and an oscilloscope should enable the electronically minded experimenter to tune the system to produce a semblance of a sine wave. For 120-volt

Table 2.1

A Comparison of Inverter Types

Type	Waveform	Main Advantages	Main Disadvantages	Cost/kW Capacity ($)	Efficiency (%)
Rotary	Sine	• Low Cost • Long Design Experience	• Moving Parts Need Maintenance	600–1200	80–90
Simple Switching Solid-State	Square	• Low Cost • Few Components • No Moving Parts	• Poor Waveform • Poor Regulation	300–600	60–90 Depends on Load
Solid-State	Sine	• No Moving Parts • Good Waveform • High Efficiency	• High Cost	1500–2000	60–80 Depends on Load
Synchronous	Sine-In Phase with Utility	• Can Feed Power Back to Utility • No Moving Parts	• Utilities Dragging Feet on Allowing Paybacks • Possible Safety Hazard to Utility Workers	900–1500	80–90 Advertised
Advanced Design Switching	Approximation of Sine	• High Power Handling Capacity • Low Cost • No Moving Parts	• Some Require Complex Storage System	500–800	85–95

incandescent lights and heating elements (toasters, irons and so on), the quality of the waveform makes no difference at all. But then, these resistance-type devices would operate just as well from 120-volt DC, and an inverter isn't even needed if the solar electric system is designed to produce that voltage. For some purposes, such as record players and tape recorders, precise 60-cycle sine waves are needed, and only the higher-quality inverters can be used. For a remote homestead, it might be best to use DC for most applications and add a small inverter for those appliances that need high-quality AC.

The new switching inverters now being made by Heart and others offer a stable waveform and high efficiencies at small loads. The no-load power drain of these designs is so low that the unit can be left on almost continuously, offering the convenience of 120-volt AC electric power outlets throughout a remote solar electric home. This makes unnecessary 12-volt versions of the usual home appliances.

Synchronous Inverters

The **synchronous inverter** is an excellent interface between a solar electric array and the available utility power. These devices, originally developed for wind energy systems but applicable to any source of DC current, convert DC to AC in synchronization with the power line. The device is simply plugged into a wall outlet and connected to the DC source. No storage batteries are needed. If the solar array is producing more power than necessary at the moment, the excess is fed back into the utility grid, which acts as the "storage" system. The utility is now required by law to purchase the excess power at some equitable rate, but they are understandably reluctant to participate in such a program. The synchronous inverters, commercially available from Gemini and others, are designed with a fail-safe relay to disconnect the inverter from the power line if the utility power fails, so that utility company workers cannot get shocked accidentally while repairing supposedly dead lines.

Since the new high-power silicon switching transistors and thyristers have become available and inexpensive, a number of researchers are developing new switching alternators that operate by turning the current delivered by solar cells or batteries on and off very quickly in a pattern that, with a small amount of filtering, closely approximates a sine wave. Aerospatiale in France has reported on such a device that efficiently produces either single- or three-phase AC from a DC source. Others have reported switching inverters that operate with efficiencies as high as 98%, depending on the load relative to the rated capacity of the device. These devices can also be run in synchronization with a utility grid.

Since solar cells are so expensive at the present time, the only justification for using this synchronous system is to demonstrate the techniques. But in the future, when solar array prices drop as expected, such systems may become commonplace. There will have to be a great deal of cooperation from the government and the utilities, however, to work out solutions to such problems as load leveling, networks, and large-scale storage.

RECOMMENDED READINGS

Cook, Stephen. Information packet. **compuSOLAR,** Gum Springs Road, Jasper, AR 72641.

Davidson, Joel. *The New Solar Electric Home: The Photovoltaics How-To Handbook* (Ann Arbor, MI: **aatec publications,** 1987).

McGeorge, John. "How to Keep Your Batteries Charged with the Sun," *Alternative Sources of Energy* 49:52–53 (May/June 1981).

Rauschenbach, Hans S. *Solar Cell Array Design Handbook* (New York: Van Nostrand Reinhold Co., 1980).

Rosenblum, Louis, *et al.* "Photovoltaic Power Systems for Rural Areas of Developing Countries," *Solar Cells* 1:65–79 (1979).

Russell, Miles. "Residential Photovoltaic Power Systems for the Northwest," *15th IEEE Photovoltaic Specialists Conference Proceedings* (1984).

Stewart, John W. *How to Make Your Own Solar Electricity* (Blue Ridge Summit, PA: TAB Books Inc., 1979).

Chapter 3
Batteries and Other Storage Systems

For most solar cell applications, some sort of storage system is needed. Many systems have been proposed or developed for the storage of electrical energy. Batteries, capacitors, flywheels, raising heavy weights, pumping water uphill, compressing air into high-pressure tanks or underground reservoirs, and even converting water to hydrogen and oxygen have all been used successfully to store electricity. These systems are reversible, so that most of the electrical energy used to charge the system can be removed when the system is run backwards. In this chapter, we will concern ourselves with storage systems suitable for the average small user of photovoltaic arrays: batteries, capacitors and mechanical storage systems.

STORAGE BATTERIES

For most solar cell applications where storage is needed, **secondary** or **storage batteries** are the best alternative. A great variety of storage batteries have been developed. About the only thing some of them have in common is the ability to be recharged. (Primary batteries such as the carbon-zinc dry cell used in flashlights cannot be completely recharged, even once, and cannot be used in any

41

practical storage system.) All these batteries, however, operate on the principle of changing electrical energy into chemical energy by means of a reversible chemical reaction.

Lately, a great deal of research and development has gone into new storage battery concepts, since electricity storage is a very important, but weak link in solar and wind energy systems, as well as in electric vehicles. Researchers have investigated such exotic battery combinations as sodium-sulfur, lithium-metal sulfides, and aluminum-air (which isn't a true storage battery since the reaction product, aluminum oxide, must be shipped back to the refiner to be smelted back into aluminum.) The most promising of these new systems is probably the zinc oxide battery developed by General Motors and Gulf & Western. Some of these batteries operate at room temperature, while others must be heated.

We will discuss the three types of storage batteries presently available: lead-acid, nickel-cadmium, and nickel-iron.

THE LEAD-ACID BATTERY

This battery is the most common type used for energy storage. The ordinary automobile battery is a lead-acid type that has been developed over the years to a design that is a compromise between low cost and reliability (for that particular use). A number of other lead-acid designs have been developed for electric vehicles, such as forklift trucks and golf carts, and as standby batteries for telephone systems and other uninterrupted power uses. A lead-acid battery designed for one particular use will not necessarily work very well in another application, so knowledge of the different kinds is necessary in order to make the proper selection.

When current is drawn from the lead-acid battery, the lead oxide positive plate is converted to lead sulfate by reacting with the sulfuric acid in the battery fluid or **electrolyte**. The negative lead plate is also converted to lead sulfate. These reactions cause electrons to flow through the external circuit from one plate to another. When the battery is recharged, current is forced back into the battery, causing the reaction to reverse and restoring the plates to their original states. In a good secondary battery system, this reversible reaction can be carried out for hundreds or thousands of cycles. (For more information on the theory and operating principles of batteries, see Recommended Readings.)

The size or **cell capacity** of storage batteries is expressed in **ampere-hours (AH)**. This is the total amount of electricity that can be drawn from a fully charged battery until it is discharged to a specified battery voltage, and is given for a specified discharge time. For example: an automobile battery of 100-AH capacity could theoretically deliver 1 ampere for 100 hours or 100 amperes for 1 hour. In actual practice, the slower the discharge rate, the greater the capacity—so the capacity rating specifies the time, 20 hours in most instances (expressed as C/20). This means that if 5 amperes were drawn for 20 hours (5 amps x 20 hr = 100 AH), the battery would be completely discharged. As we will see later when battery capacity is discussed in more detail, the actual usable battery storage capacity is less than that figure. The temperature and other factors also affect the usable battery storage capacity.

The output voltage of each individual cell of a lead-acid battery is dependent on a number of factors, *e.g.*, temperature and state-of-charge, but for the sake of categorizing batteries, it will be taken as 2 volts. The total battery voltage is the sum of the individual cell voltages, so a battery with 3 cells would be called a 6-volt battery while a 12-volt battery has 6 cells in a series. Batteries are normally available in 6-, 12- or 24-volt output, but some tractor batteries are rated at 8 volts.

First, let us consider the automobile battery, since it is the most familiar and the easiest to acquire. Automobile batteries are designed to deliver a very large current for a short time in order to operate the starter. Normally, they are not discharged more than a few percent of their capacity. The lead plates, or positive and negative electrodes, are paper-thin so that a great number can fit into a small space. This large active surface area in the plates allows currents of 200 amps or more to be drawn for a few seconds at a time without damage to the battery. Once the car is started, the alternator furnishes all the electrical requirements of the automobile and recharges the battery for the next start. If the automobile's electrical system is working properly, the battery is no longer used.

If the alternator fails and does not recharge the battery, or if you leave the headlights on overnight, the battery will become completely discharged. The battery can usually be recharged and will continue to work properly, but a small amount of permanent damage is done each time this happens. The normal automobile battery is designed to take 20 or so of these **deep-discharge cycles** before becoming completely useless. The battery is also designed to last anywhere from 2 to 5 years of normal automobile use, depending on the quality of the battery. This is the same as the guarantee period (5 years is the longest expected life, even for "lifetime" batteries).

In newer automobile batteries the cells are sealed, with only a tiny vent hole and no caps for adding water. Two technical changes make this type of battery possible: first a lead-calcium alloy is used in the plates to make them more stable and, second, a catalyst is placed in the top of the cell to react with the gas released during charging to convert it back into water which drips back into the cell. These "no-maintenance" batteries normally have no provision to get into the cells, but some high-quality batteries have removable caps set flush and hidden under the plastic information sheet glued to the top of the battery. Older batteries were made with black hard rubber cases, but newer cases are polypropylene and some are translucent so that the fluid level in the cells can be checked visually. The type or color of the case has no effect on battery performance. If a set of automobile batteries is used with a solar array, a long cloudy spell could discharge them to 75% or 80%—the same as leaving the car headlights on—and only a few such discharges would destroy the storage system. Thus, for use with solar cells, a different type of battery is required.

The Deep-Discharge Lead-Acid Battery

There are a number of applications, besides solar cells, which require a deep-discharge battery. A cruising sailboat will, for example, use small amounts of electricity over a period of days for running lights and navigation equipment before it is possible to recharge the battery. (This is one of the reasons why solar cell arrays are so useful on sailboats.) Electric trolling motors also need this kind of battery. A **marine battery** with this deep-discharge capability will work quite well with a small solar cell array.

Forklift trucks and golf carts use large-capacity deep-discharge batteries which are designed for long life and many discharge cycles. Cycle lives of 1000 to 2000 cycles are typical of these **motive-power batteries**, which are expected to last 15 years or more. These batteries are heavy and expensive since they are built with thick lead plates and a greater electrolyte capacity to make them more reliable under difficult operating conditions. A set of these batteries, if acquired at a decent price, could give years of satisfactory service at low total cost in a solar cell application.

Another common application for lead-acid batteries is in emergency standby power systems. These **stationary batteries** are kept fully charged by a regulated charging system which furnishes all the current normally needed by the system. Only in the event of a power outage is the battery actually used; the rest of the time it is "floating" in the circuit. These batteries are designed to have a very long life under these conditions (25 years or more), but a poor deep-discharge cycle life.

Finally, in recent years, manufacturers have begun making batteries specifically designed to be used with photovoltaic power systems. An example of such a battery is the Delco 2000 Photovoltaic Battery, but similar batteries are made by other manufacturers, such as Gould and Exide. Table 3.1 gives a summary of the characteristics of lead-acid batteries intended for different services.

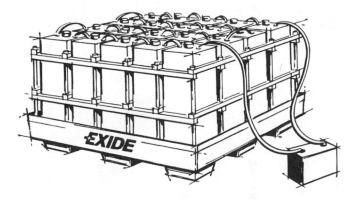

FIGURE 3.1–Deep-cycle battery, Exide Renewable Energy Series PHv-DL.
(Courtesy of Exide Corporation, Horsham, Pennsylvania)

Table 3.1
Lead-Acid Battery Types

Type	General Characteristics	Typical Applications
Automotive (SLI)	High discharge rate, relatively low cost, poor cycle life	Automobile starting, lighting and ignition; tractors, snowmobiles and other small engine starting
Diesel Starting	High discharge rate, more rugged than automotive, more expensive	Large diesel engine starting; can be used for small photovoltaic systems
Motive Power (Traction)	Moderate discharge rate, good cycle life	Fork lifts; mine vehicles; golf carts; submarines; other electric vehicles
Stationary (Float)	Medium discharge rate, good life (years); some types have low self-discharge rates, poor cycle life	Telephone power supplies; uninterruptible power supplies (UPS); other standby and emergency power supply applications
Sealed	No maintenance, moderate rate, poor cycle life	Lanterns, portable tools, portable electronic equipment; also sealed SLI
Low-Rate Photovoltaic	Low maintenance, low self-discharge, special designs for high and low ambient temperatures, poor deep cycle life	Remote, daily shallow discharge, large reserve (stand-alone) photovoltaic power systems
Medium-Rate Photovoltaic	Moderate discharge rate, good cycle life, low maintenance	Photovoltaic power systems with onsite backup or utility interface, requiring frequent deep-cycle operation

Lead-Acid Cell Characteristics

There are three important characteristics of lead-acid batteries that must be considered when designing and sizing solar cell array battery storage systems. First, the voltage output of a battery is a function of temperature and state-of-charge. Second, the useful capacity of the battery decreases significantly with a decrease in temperature, and, third, batteries will slowly discharge on standing. This **self-discharge rate** is also a function of temperature and battery design.

Output Voltage

Figure 3.2 shows the output voltage of a typical lead-acid cell, and how it varies with the depth of discharge and the rate of discharge. At the discharge rates expected from a photovoltaic storage battery system, the curve will resemble the line marked C/10 (that is, the 10-hour capacity rate). The charging voltage curve of the battery is similar but at a slightly higher voltage. Figure 3.3 is a typical solar cell I-V curve but plotted inversely to those shown in the earlier chapters. The two curves are remarkably similar. This similarity makes it possible, in relatively small systems, to connect the solar cell array directly to the storage battery without the need of a voltage regulator. Only a blocking diode is needed to ensure that the battery cannot discharge back through the solar cells when the array voltage is lower than the battery voltage. As the array charges the battery, the voltage of the system rises and the current output of the array drops off to that necessary to trickle-charge the battery. This system works well with 12-volt batteries if the array has 32 silicon cells in series and if the rate output of the cells is between 0.6 and 1.5 amps for every 100 AH of storage battery capacity.

This simple system will work well as long as the battery and array temperature do not vary too far from room temperature. This is true of the sailboat battery charger, for example, where the battery is inside the boat and will stay at a relatively constant temperature and the solar cell array will never be very cold while it is operating. If the solar cell array gets very warm, it will decrease

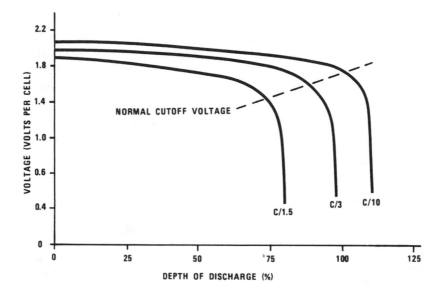

FIGURE 3.2–Output voltage of typical lead-acid battery at different
states of charge and discharge rates.

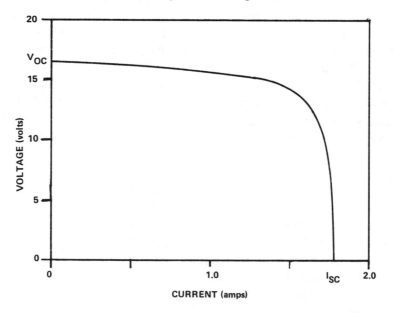

FIGURE 3.3–Solar cell module characteristics.

the charging efficiency of the system but will do no harm to any of the components.

Figure 3.4 shows the effect of temperature on the output voltage of a 32-cell module and on the voltage needed to charge a nominal 12-volt storage battery. Assume there is a blocking diode in series with the array. The outputs of both solar cells and batteries drop when the temperature goes up, but they vary at different rates. At very cold temperatures it is possible that the output of the solar array could increase to the point where it is overcharging the battery, particularly if the battery is indoors or protected against temperature extremes. This overcharging will, at the least, cause gassing and excessive loss of water but could cause the active material of the positive plate to loosen and flake off, reducing the capacity and expected life of the battery. Normally, the sun striking the solar array will heat it, minimizing the increase of output voltage with increasing sunlight intensity. In small systems where the battery capacity is relatively large compared to the amp or so of maximum output current from the array, there should be no problem. But in larger systems where the output current of the solar array may be large compared to the trickle-charge rate needed for the set of storage batteries, overcharging could occur under certain circumstances in cold weather, and a voltage regu-

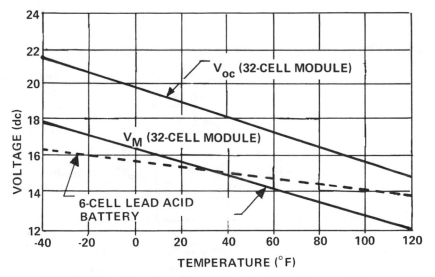

FIGURE 3.4—Module and battery temperature characteristics.

lator might be needed. These regulators are commercially available, specifically designed for use with solar arrays, with complete temperature compensation built in. Or the simple shunt regulator described in Figure 3.5 could be built. This regulator ensures against overcharging but it does not reduce the efficiency of the solar cell system. If such a regulator is used, a 33- to 36-cell array might be appropriate to achieve a greater charging voltage under cloudy conditions.

Storage Capacity

The amount of storage battery capacity needed is determined by the expected load and the longest time over which the system is expected to operate on the storage batteries alone. (See the section on calculating load requirements in Chapter 2 to find out how to determine the expected load.) Multiply the daily ampere-hour requirements by the number of days of storage needed. In actual practice, 5 to 7 days of storage is adequate to account for deviations in the weather pattern, particularly if you are at the installation and can cut back on energy consumption if the batteries get too low. For an unattended, inaccessible site where no backup is available, 8 to 10 days of storage capacity may be

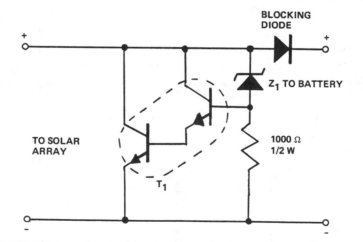

FIGURE 3.5—15-Volt shunt regulator with 20 amps maximum current handling capacity T_1 is 2n6282 npn power darlington. Transistor Z_1 is IN5352 15-volt Zener diode.

required and high-quality batteries with low internal leakage should be used.

Using the example of the cruising sailboat with a determined need for 30 AH of power a day, and which has an auxiliary engine with an alternator to recharge the system when under power, 5 days storage capacity should be more than adequate to meet the requirements: 5 days x 30 AH = 150 AH of battery capacity needed for the boat's electrical system. This would be two 75-AH marine batteries. It is important to realize that, for safety, a third separate battery should be used for the auxiliary engine's starter. However, the solar panel can be connected to this starter battery through a separate blocking diode to ensure that the battery stays fully charged, even if the engine isn't started for days at a time.

One important reason for having adequate battery storage capacity is that the expected life of the battery is greatly influenced by the depth of discharge in the cycles. Figure 3.6 shows this relationship for two types of batteries. In actual practice it is best not to discharge batteries more than 60% or so, and this only on occasion. If the average discharge is less than 30%, which is what you would expect using the guidelines given above, the batteries should last their design lifetime.

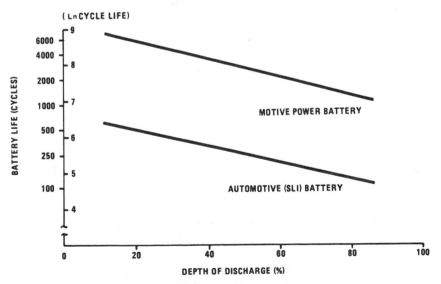

FIGURE 3.6—Effect of depth of discharge on lead-acid battery cycle life.

The available battery capacity is a function of the discharge rate. Figure 3.7 shows this relationship for a number of different battery types. The larger the battery storage capacity, the less the drain on each battery for a given load current. This will make the reserve proportionately larger. Doubling the number of storage batteries will more than double the capacity.

One factor which greatly reduces available storage capacity is cold weather. Figure 3.8 shows battery capacity at different temperatures. At temperatures below freezing the capacity drops off sharply. (This is one reason why cars are hard to start in the winter.) Also, a battery can freeze in the winter unless it is well-charged, since the electrolyte in a discharged battery is essentially water. So it is best, when possible, to store batteries in a place that will not get too cold.

Self-Discharge

Finally, batteries will slowly discharge while standing. Figure 3.9 shows how this discharge rate varies with the temperature and the composition of the plates. Most automobile batteries have lead-antimony plates, but the new low-maintenance sealed bat-

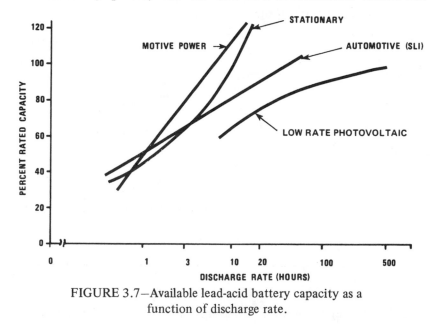

FIGURE 3.7—Available lead-acid battery capacity as a
function of discharge rate.

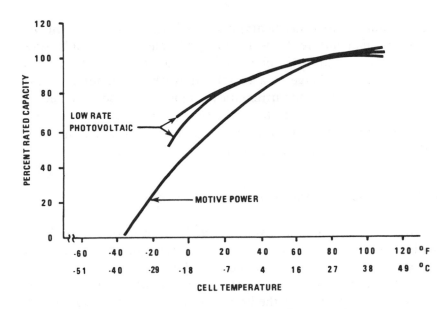

FIGURE 3.8–Lead-acid battery capacity as a function of temperature.

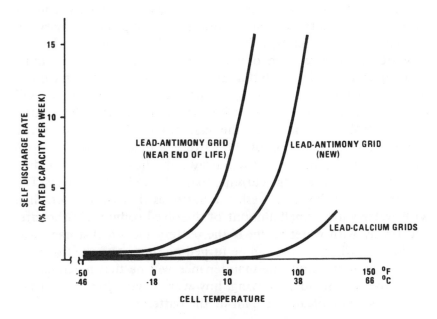

FIGURE 3.9–Lead-acid battery self-discharge rate.

teries normally use lead-calcium plates. As the curve clearly shows, at room temperature lead-calcium grids have a negligible self-discharge rate while the older lead-antimony grid batteries can lose their entire charge in a couple of weeks of standing. These batteries may even leak current faster than a small solar cell array can furnish it and the system will slowly lose charge even on a series of sunny days. Obviously, if long-term storage is needed, batteries with lead-calcium (or pure lead) plates must be used.

A Practical Battery Storage System

From the information given above, it is clear that using lead-acid storage batteries is more complicated than placing a collection of batteries somewhere and connecting the wires.

First, the batteries must be stored in the proper environment. The best plan is to place the batteries in a rack so they are off the floor and accessible for cleaning and maintenance. Since batteries contain sulfuric acid and give off a fine corrosive mist during charging, it is important to protect both the racks themselves and the wiring used to connect the batteries against corrosion. The gas which bubbles up from a battery during the last stages of charging, particularly if the battery is being overcharged, is an explosive mixture of hydrogen and oxygen. The battery storage area must be adequately ventilated so that an explosive concentration cannot build up. If the battery has the proper vent caps, or is a "sealed" type, a chamber inside the vents is supposed to separate the sulfuric acid electrolyte liquid from the gas, but inevitably some spray escapes and coats the top of the battery with an extremely corrosive slime. This must be washed away periodically as it is conductive and can cause leakage currents to discharge in the battery. The best way to do this is to wash the batteries first with water, then with water with a small amount of dissolved sodium bicarbonate, and finally with water again. If the battery racks and storage area are located where a hose can be turned on them, and the floor can get wet without harm, the maintenance will be that much easier. Don't overdo the hose spraying, however, as you do not want tap water or sodium bicarbonate inside the batteries.

The cables that interconnect the battery terminals should be heavy and the connections very tight. Automobile batteries have tapered post terminals; connectors for these are available at any auto supply store. Most other storage batteries, however, have a different style terminal with a threaded stud. This stud is usually copper- or lead-plated and copper or lead end terminals should be used. (I have used stainless steel nuts to fasten the connectors, but copper- or lead-plated nuts also work well.) Steel nuts will corrode very quickly under these conditions. Any terminal resistance will require a higher charging voltage and waste power in the discharge cycle. The exact wiring system will depend on the battery voltage compared to the solar array voltage. This topic is covered in Chapter 2.

Maintenance

Regular maintenance of the battery system is essential if it is to last. In addition to cleaning, the condition of the batteries must be checked at regular intervals, and water added occasionally to all but the sealed batteries. Only distilled water should be used. Regular tap water contains small amounts of chemicals, *e.g.,* calcium and chlorine, which are bad for battery chemistry. Also, no "rejuvenator" chemicals should be used. If a battery cannot be rejuvenated by a proper charging sequence, it probably needs to be replaced.

A good hydrometer is necessary to check the exact charge condition of each cell of a battery. The better ones have an actual float instead of a series of balls, and a built-in thermometer with tables, so the state-of-charge at different temperatures can be determined. If the battery caps are not removable and you cannot use a hydrometer, the state-of-charge can be estimated by measuring the battery output voltage when a small current (1 amp per battery is sufficient) is being drawn. Figure 3.2 can be used to estimate the condition of the battery (using the C/10 curve). New Delco batteries come with an "eye" that tells the condition of at least one cell of the battery. Quite often, only a single cell of a battery will fail prematurely and chances are it won't be the cell with the indicator.

Sulfation

The worst thing that can happen to a battery is to sit completely discharged and unused for any length of time. Sulfation—a condition where large crystals of lead sulfate grow on the plates in place of the tiny crystals normally present—will make the battery extremely difficult to recharge. Even a partially discharged battery that isn't used for a period of weeks will show evidence of this. The causes and cures of this phenomenon have not been studied adequately, but sulfation seems to take place more readily at higher temperatures and may be partially reversed by a carefully controlled recharging of the battery. Start the recharging at a low current level (less than 2 amps for a 0.75-AH battery), and as the cell resistance decreases, as evidenced by a *decrease* in the charging voltage needed to maintain a constant current, increase the charging current to 10 or 15 amps until significant gassing occurs. This bubble formation will help stir the electrolyte, bringing fresh solution into contact with the plates. **REMEMBER: The gas given off is extremely explosive, so make sure there is adequate ventilation and keep sparks and flames away from the battery**. If it is possible to measure the specific gravity of the battery, check all cells and continue this overcharging condition until all cells read the same. The reading should be above 1.25 (or whatever represents a full charge for the particular battery).

Some of the plate material may become dislodged and flake off during this process and the battery may never have the original ampere-hour capacity, but at least the sulfated battery will be restored to usefulness. Very badly sulfated batteries may not respond to this treatment and will have to be traded in for their scrap value.

Used Batteries

Quite often it is possible to find a set of used batteries appropriate for a solar cell storage system. If it is possible to purchase them for the scrap value of the lead, no money is lost if the batteries turn out to be unsuitable, since you can still trade them in to recoup your investment. It is best to remember that organizations that use a large number of batteries usually have maintenance personnel who know how to care for them and generally will not get rid of them unless they are pretty well finished. However, batteries that are no longer reliable for severe service, in forklift trucks or golf carts, can still last for years in a solar array system, if they are charged and discharged slowly and never allowed to sit for any length of time almost completely discharged.

Another source of satisfactory used batteries is dealers who maintain and replace diesel truck batteries for fleet owners. These diesel starting batteries, while not true deep-discharge batteries, are built with much thicker plates and longer design lives than the best automobile batteries. When used in a system where they are discharged slowly, and rarely below 50% of capacity, they will last for a surprisingly long time. It is sometimes possible to purchase an identical set of newly removed batteries for little more than the scrap value, and since large fleet owners replace batteries on a calendar basis, regardless of use, such a set may give three or more years of satisfactory operation before they need to be replaced. Try to get a set with lead-cadmium or pure lead plates.

Stationary or float batteries that have been taken out of service by the telephone company, or from standby power units, can also be used. Sometimes these batteries can be disassembled for cleaning and repair. It is best to find an expert to perform this work, as the procedures are complex and dangerous.

Wind energy system dealers generally know a great deal about batteries, particularly used ones, since there were once thousands of wind generators with battery storage systems on farms all over the country. One of these battery sets, even 50 years old, if renovated and treated with proper care, can last another lifetime in a solar electric system.

Used automobile batteries should be tried only as a last resort or as a temporary measure until the proper battery can be found. A so-called "renovated" automobile battery is usually nothing more than a traded-in battery that has been cleaned, repainted and charged. Since many people buy a new battery when the real problem with their car is some other electrical problem, the traded-in battery may have a lot of life left.

NICKEL-CADMIUM BATTERIES

These storage batteries, which were first developed after the turn of the century, were not commonly used in the United States until the 1950s. The two plates of this battery are a nickel plate packed with nickel oxide and a cadmium plate. The main advantages of the nickel-cadmium (ni-cad) battery are long life and reduced maintenance requirements. The main disadvantage is the high cost per ampere-hour of capacity. In smaller systems, the convenience of a sealed battery semipermanently installed in a device outweighs this higher cost. Most rechargeable flashlights, power tools and pocket calculators have some sort of nickel-cadmium battery.

Ni-cad batteries are manufactured in two basic types: "sealed" and vented. The sealed type actually has a pressure relief valve built into the cell that releases to prevent an explosion if the cell is heated or greatly overcharged. Some of these valves will reclose after they pop open. Others stay open allowing the water to evaporate from the electrolyte, quickly ruining the battery. The vented type have resealable vents that open and close under small pressure changes. The electrolyte, a solution of potassium hydroxide in water, is used only to carry the current between the plates, and does not change when the cell is charged or discharged. The nickel oxide electrode is a sintered plate or pocket plate construction battery with the sintered plate being the one most used in small cells, of the sealed design.

The voltage output of a ni-cad battery is 1.2 volts per cell and changes very little with use until the battery is almost completely discharged; then it will drop sharply. A cutoff voltage of 1.0 volt is usually used as an indication that the battery is discharged. A ni-cad battery can accept a charge at a relatively high rate (C/1) and is capable of operation under continuous overcharge provided that the charging current does not exceed a given amount (C/15). To give a numerical example: A nickel-cadmium sealed cell with an ampere-hour capacity of 15 could initially stand a charging current of up to 15 amps, but as the cell becomes charged, this would have to be reduced to a final current of 1 amp. This 1-amp current could run through the battery indefinitely without harming it. Of course, the battery could also have been charged at a steady rate of 1 amp from the beginning.

Nickel-cadmium batteries can be deep-discharged many times without harm and have a much smaller change in performance with temperature compared to the lead-acid battery. All these properties make them ideal storage batteries to use with solar cell systems where the battery bank can be connected directly to the solar array with no need for a voltage regulator. (A blocking diode would still be needed, however, to keep the array from discharging the batteries in the dark.) It is important that the maximum output current of the array does not exceed the C/15 continuous overcharge current of the battery.

There is one quirk of the ni-cad battery that needs mention. This is the memory effect. For example, after a ni-cad battery has been repeatedly discharged by 25%, the cell voltage drops off sharply at this state-of-discharge, and the battery acts as if it has only one-quarter of its true capacity. This effect is shown in Figure 3.10 together with the voltage versus discharge characteristics of a normal cell. Prolonged periods of overcharge will also produce such an effect. In most cases, the battery can be restored to full capacity by letting it discharge until it is completely dead and recharging it completely, but without allowing it to over-

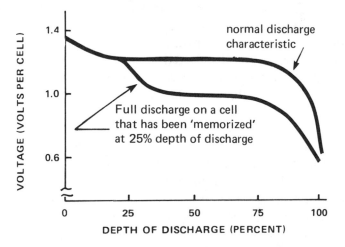

FIGURE 3.10—Memory effect on discharge voltage of sintered plate nickel-cadmium battery.

charge. This trick will work two or three times on a ni-cad battery before permanent damage is done.

In conclusion, ni-cad batteries are very useful for small solar-powered systems but too costly for larger systems unless the increased reliability and low maintenance requirements are worth the premium price. This is the case for critical equipment at a remote or inaccessible site like a Coast Guard buoy or a space satellite power system.

NICKEL-IRON BATTERIES

These batteries were developed by Thomas Edison in the late 19th century as a more reliable alternative to the lead-acid battery, and were used extensively by the railroads as storage batteries on passenger cars. They are extremely rugged devices, not bothered by the abuse that would completely destroy the usual lead-acid battery. However, they have some electrical disadvantages. The first, not greatly important for stationary use, is that the output voltage is low, about 1.1 volt per cell, so that approximately twice as many cells are needed to produce the same battery voltage. This, coupled with the relatively large plate size needed for a given storage capacity, produces a much larger and heavier battery for the same ampere-hour of capacity. The second disadvantage is that the output voltage drops more quickly as the battery is discharged, so that a half-discharged nickel-iron battery has a much lower voltage than a half-discharged lead-acid battery. If the battery were used in a float system where, under normal operation, the state-of-charge changes slowly and over a small range, the charge in output voltage would be minimized. Another solution is to design the photovoltaic and storage system to produce an excess output voltage when fully charged and use a voltage regulator to control the final output voltage.

A more practical problem with Edison batteries is finding them. They are currently not being manufactured, although surplus dealers may have old units removed from passenger cars. If you can find a set of these batteries in good operating condition and can locate someone who is familiar with their care and maintenance, you should end up with a first-class electrical storage system at a very low cost.

SMALL STORAGE SYSTEMS

For digital watches and pocket calculators, small, so-called rechargeable batteries are sometimes used. In these micropower systems, the distinction between primary and rechargeable batteries becomes lost as, at a very slow rate, any battery action is reversible to an extent. Many of the "solar" digital watches use the same silver oxide or mercury battery that their nonsolar counterparts use and rely on the solar cells to extend the normal lifetime of the batteries, which is measured in years in any event. Work on a truly rechargeable tiny battery is now under way in Switzerland and Japan, as well as in the United States. The most interesting concept is a solid ceramic battery in which diffusion of small ions through a crystal or amorphous matrix replaces the liquid electrolyte. Such batteries should have useful lifetimes, in watches, measured in decades.

For very small loads, a capacitor may serve to store small amounts of current for a short time and to smooth short-term changes in the load or solar cell output. For example, a 4000–8000-microfarad electrolytic capacitor can aid in delivering the starting current for a small motor when the running current is furnished entirely by solar cells. A capacitor can also be used to ensure a continuous supply of power to a solar-powered calculator when there is no battery or on/off switch. This will prevent a momentary shadow from shutting off the integrated circuit and erasing the calculation in progress.

MECHANICAL STORAGE SYSTEMS

Most mechanical storage systems developed to store electricity are either too complicated or are applicable only to large-scale use. Nonetheless, there are situations where mechanical storage can be very useful, even in small systems. These are situations where the electricity is just a medium and the desired end product is stored. One example is a water pump that runs while the sun shines, pumping water into a reservoir, to be drawn upon as needed. Another example is a solar-powered refrigerator that has holding plates that can keep food cold for two or three days when completely charged by the photovoltaic-powered electric compressor. This idea could be extended to air compressors with large storage tanks and similar systems. A simple diode or relay switching system could be used to direct the solar cell arrays' output to a set of storage batteries when the system is charged or filled to capacity.

Flywheels have been proposed as small- and medium-sized storage systems, and have actually been used in some European homes for years, but considerable development needs to be done before they will be available in easy-to-use, packaged systems. It has been calculated that flywheel systems, using space age materials like boron fiber composites, should be able to effectively store two to three times the energy as the equivalent weight in currently available storage batteries.

HYDROGEN

It is very simple to use the low-voltage DC output from a series string of three or four solar cells to separate water into hydrogen and oxygen. Put a pair of electrodes into the water and connect them to the solar array. A few drops of sulfuric acid in the water will make it sufficiently conductive to pass the necessary current. This system will utilize 10 to 12% of the energy in sunlight to make hydrogen.

It is also possible to produce hydrogen directly in a photoelectrochemical cell, but the efficiency of the best of the experimental systems developed so far falls short of the simple indirect system described above. This is because, in the indirect system, the materials used in the light absorption and hydrogen evolution steps can be optimized for their particular functions.

There has been a great deal written about the **hydrogen economy**, a system in which hydrogen is substituted for fossil fuels in many technological applications in transportation and industry. There is one important point to remember: hydrogen is not a primary energy source, but rather a way of storing energy. All the energy that can be generated by using hydrogen represents energy that was put into the hydrogen when it was electrolyzed from water. Because of inefficiencies in the process, more energy (from some other source) was used in making the hydrogen than can be extracted later from the hydrogen.

The energy stored in hydrogen can be used in a number of ways. Hydrogen can be burned like natural gas to furnish heat, it can be used as a fuel in a piston engine or turbine, it can be oxidized in a fuel cell to generate electricity, or it can even be burned in a rocket engine to power a space shuttle. Thus, using sunlight to produce hydrogen leads to a very flexible set of options. However, hydrogen made in this fashion is expensive and represents a good deal of high-quality electrical energy. It would be much cheaper and more efficient, for example, to use a solar water heater and store the heated water than to produce hydrogen with a photovoltaic system and later burn the gas to heat water.

The hydrogen economy discussed by some authors is conceived of as a large-scale network of hydrogen producers, pipelines, and users similar to our present natural gas utility system—the important difference being that hydrogen is a renewable source of energy whereas natural gas is a fossil fuel in limited supply. Presently, most natural gas is used to produce heat that could be furnished directly by the sun through relatively simple systems. The hydrogen economy would make the most impact in the future in systems which need more than low-temperature heat. The hydrogen can be converted to electricity efficiently using the fuel cells developed for the space program, but it is easier and more logical to transport and store electricity than hydrogen.

Hydrogen is also a very clean-burning fuel that could be used in automobiles or other transportation systems. However, it is a very light gas and would have to be condensed in volume before a vehicle could carry sufficient fuel. The energy required to compress a volume of hydrogen to the 300 pounds or so necessary to fill a tank of reasonable size—so it can power the vehicle 200 miles or more—or the energy needed to cool the gas and produce liquid hydrogen (which also occupies a much smaller storage volume) is an appreciable fraction of the energy that will be released when the hydrogen is burned. Of course, some of this energy might be recovered in an expansion engine on the vehicle. However, the best way of storing hydrogen in a small volume is to react it chemically to form a liquid or solid product. Metal hydride systems are being developed that soak up and release large quantities of hydrogen with small pressure changes at room temperature. Calculations indicate that such systems could carry enough hydrogen in a small volume to make them practical for transportation systems.

Hydrogen can also be reacted with carbon dioxide from the air to make solid or liquid organic compounds such as alcohols. These can be burned as fuel or used as starting materials for a wide variety of products including food, drugs and plastics. The best known and most completely developed such system is photosynthesis by plants. Here the energy of sunlight is used to separate hydrogen from water and combine it with carbon dioxide to form sugars. Later, the sugars are rearranged and assembled into an array of substances. Carbohydrates or sugars represent a very convenient way of storing the energy from sunlight. In a sense, all life depends on this hydrogen economy.

RECOMMENDED READINGS

"Battery Service Manual," $2, The Battery Council International, 111 East Wacker Drive, Chicago, IL 60601.

Bockris, J. O'M. *Energy. The Solar-Hydrogen Alternative* (New York: John Wiley & Sons, 1975).

"Capturing the Sun—Batteries and Solar Energy Storage," Lead Industries Association, Inc., 292 Madison Avenue, New York, NY 10017.

Handbook for Battery Energy Storage in Photovoltaic Power Systems (San Francisco: Bechtel National, Inc., Research and Engineering Operation, 1980).

Vinal, George. *Storage Batteries,* 4th edition (New York: Wiley-Interscience, 1955).

Chapter 4
Assembling
Your Own Solar Array

TESTING SOLAR CELLS

There are a number of sources from which the average person can obtain individual solar cells. (Appendix A lists names and addresses.) These cells range from well-tested and calibrated devices guaranteed to perform to their given specifications, through standard production quality cells, to untested or reject cells that can be purchased at a substantial discount. Even if you purchase a standard guaranteed cell, it is important to test each cell thoroughly before using it in a device or an array. The amount of work entailed in removing one cell from a finished array and the danger of ruining the neighboring, good cells make testing individual cells a must.

Fortunately, testing solar cells is neither difficult nor complicated. The apparatus needed can be as simple as an inexpensive multimeter or voltmeter and ammeter of the proper range. Electronic supply houses sell these meters for $10 and up. A fancy digital meter is not needed, but it is good to have one which has, among its selection of ranges, a voltage range as low as 1 volt or less and a current range above 2.5 amps. For testing completed arrays, we will be measuring voltages of 16 to 18 volts or more, and short-circuit currents possibly as high as 2.5 amps, depending on the size of the array and the light conditions.

The simplest test which gives meaningful information measures a cell's short-circuit current and open-circuit voltage under bright sunlight conditions. If bright sunlight is not available, it is possible to perform the tests outside or with the light through a window on a cloudy day or even under artificial illumination. To get meaningful results under these conditions, it is necessary to have available a standard cell of known output to measure the actual light intensity. Even then, if the light used in testing the cells is very different in color from that of real sunlight, any difference between the spectral response of the cells to be tested and that of the standard cell will produce an error. For example, on a cloudy day, when the illumination is much bluer and "colder," a blue-sensitive cell will respond more efficiently than it would to "warm" incandescent illumination of the same actual intensity. However, if the standard cell is made of the same material as the cells to be tested, these differences will be negligible. What we aim for is not a set of 32 perfect cells, but rather a matched set of cells of roughly equivalent quality, even if that quality is relatively poor compared to the best cells available.

A Simple Solar Simulator

To test solar cells indoors, you need a solar simulator. Usually, solar simulators are expensive devices with xenon lights and filters, but a very simple way to produce a reasonable simulation of AM1 and AM2 sunlight is to use a standard EHL projector lamp. These lamps, which are used in slide projectors, have special reflectors designed to let the infrared light escape from the bulb and reflect only the visible light onto the slide to keep the slide cool. By coincidence, the spectral output of this lamp is very close to that of sunlight AM1 conditions if the lamp is operated at 117 AC. If the lamp voltage is reduced to 100 volts, the output becomes redder and nearly matches AM2 sunlight. The exact distance between the bulb and the solar cell will vary with each bulb, but it is about 35 cm. You can determine the exact distance with a solar cell that has been previously calibrated, but even without this the simulator will give a constant light output so cells can be checked relative to each other.

A Simple Sample Holder

To measure the electrical properties of solar cells, it is necessary to make good electrical contacts to both the back contact and the top fingers in such a way that the cell is not shadowed. Figure 4.1 shows a simple sample holder that can be built to accomplish this. The base is a piece of plywood about 6 x 8 inches onto which a piece of thin copper or galvanized steel is glued with contact cement. The five-way binding posts, available at most electronic supply stores, serve to connect the wires from the multimeter or voltmeter and ammeter to the holder. After drilling a pair of small holes 3/4 inch apart for the shafts of the binding posts, drill a

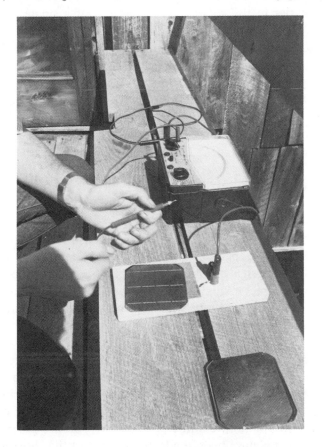

FIGURE 4.1—A simple sample holder (David Ross Stevens photo).

larger hole from the back for each part so that the nuts holding the posts in place can be recessed, allowing the plywood base to sit flat. When a cell is to be tested, the back contact is established by the cell resting on the metal plate, so the plate has to be kept clean and free of insulating oxides. The slight pressure of the front contact against the fingers should be enough to make a good back contact.

Two types of front contacts are possible for this holder. The simplest is just one of the probes that comes with a multimeter. It is the easiest to use if the holder is kept flat and there are a lot of cells to test, since a cell can simply be laid on the holder plate and the cell touched with the probe. The other type, a spring-style holder, is easier to use for intensive testings of a single cell, for it frees the hands and holds the cell securely in one place so that the holder can be tipped to face the sun directly. Also, since the pressure is constant, variations in the measured short-circuit current, caused by poor contact, are minimized.

More elaborate sample holders with temperature-controlled bases and adjustable top contacts can be constructed for more advanced research into the properties of semiconductor solar cells. But for testing purposes, the holder shown should be adequate.

CURRENT–VOLTAGE MEASUREMENTS

All commercially available silicon solar cells produce nearly the same voltage. The open-circuit voltage expected, even for relatively inefficient cells, should be over 0.5 volt and will not exceed 0.6 volt under direct sunlight illumination. This voltage will change only slightly with changes in the brightness of the sunlight. Under reduced illumination, the open-circuit voltage will drop to 0.3 volt or so.

The occasional defective cell with little or no voltage output probably is shorted. If you examine the cell and find no obvious defect, like a spot where the front contact finger has accidentally been extended to the edge where it can touch the back contact, the short is internal and cannot be fixed. If you are positive that the measurements have been made correctly and the short persists, it is possible to rescue part of the cell by deliberately breaking the cell in half, retesting, and using the good half. If the short is a point defect, the rest of the cell area will be perfectly good. Of course, the bad half can also be broken since each part of a solar cell, no matter how small, is a complete solar cell. With some clever jumper wires between the fingers, broken pieces of cells can be assembled into usable arrays.

The current output of a solar cell is directly related to the area of the cell and the intensity of the light falling onto the cell. It is also dependent on the quality of the cell and should be the main determinant of whether or not to use a particular cell in an array, since the maximum current output of a series string of cells will be essentially the output of the worst cell in the string. Table 1.1 in Chapter 1 gives the expected current output of a number of commonly available sizes of solar cells. The rated, or AM1, output of a cell is the current expected from sunlight coming from directly overhead through a dry atmosphere (the Sahara Desert at high noon) and is the one most often used in literature and specifications. The AM2 output is closer to what would normally be expected from measurements made outdoors under normal weather conditions. For silicon cells not listed in the table, a figure of 27 mA/cm² for AM1 or 21 mA/cm² for AM2 conditions could be used. Simply calculate the surface area of the cells (including the fingers) in square centimeters and multiply by those figures to calculate the expected current output. If a standardized cell is available, it can be used to measure the exact light intensity at any given moment and greatly simplify checking of the cells. You must make sure, though, that the standard and the cell to be tested are exposed to exactly the same light and are oriented at the same angle.

HOW CELLS ARE CONNECTED

Solar cells can be connected to an external circuit in a number of ways: they can be fitted into a holder with spring contacts that press onto the front and back; the connections can be spot-welded; or the circuit they power can even be built onto the same chip of silicon as the cell area. But the most common way to connect solar cells is by soldering jumpers to them.

Solar cells can be purchased with a current rating of 2.5 amps, but the half-volt output of a cell is usually too little for practical systems. The whole purpose of assembling cells into arrays is to increase the voltage or current available compared to the individual cell output. Usually, the cell is sized to deliver the required current, and these cells are connected in series to produce the required voltage. Even if the current output desired is that of two or more individual cells, it is probably better to make these up as a set of two or more separate modules and then connect the modules to the external circuit. A blocking diode should always be used in connecting an array to the battery through the array at night or any other time when the array is putting out less than the battery voltage. If a sepai ite blocking diode is put in series with each module in an array, the failure of one module will not affect the other. Figure 4.2 shows several methods that have been used to build a series string of solar cells.

Early in the space program, when solar cells were used to power the first satellites, the usual practice for assembling a series string of cells was called **shingling**. This is shown in Figure 4.2a. The bottom edge of one cell overlapped about 1 mm of the top edge of the next cell and was soldered. The finger pattern was designed with a soldering strip on one edge. However, expansion and contraction with temperature, along with any flexing, places an

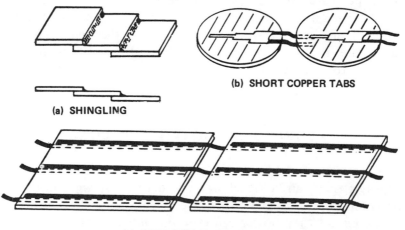

(a) SHINGLING

(b) SHORT COPPER TABS

(c) LONG COPPER STRIPS

FIGURE 4.2—Several ways of connecting solar cells in series.

enormous stress on the solder junction, and premature failure was one result.

At the present time the most common way to interconnect solar cells in an array is to use jumpers made of copper foil. Commercial solar arrays have specially shaped foil pieces preplated with tin or solder, but straight strips can be used with most finger patterns.

It is quite easy to prepare jumper strips out of copper foil; the series connection methods are illustrated in Figure 4.2 b & c. The copper foil should be quite thin (0.002 inch or 50 microns) in order to allow flexible connections that will put little stress on the cells themselves. These strips are very easy to cut from a wide piece of copper foil (see Figure 4.3), or the foil used to assemble stained

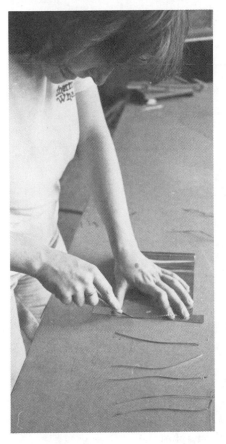

FIGURE 4.3–Preparing strips of foil
(David Ross Stevens photo).

glass pieces can be used. This foil is available at many hobby shops or stained glass supply dealers. The exact size and shape of the strips will depend on the finger pattern of the cells used, but it is best to allow about 1 to 2 mm space between cells. It is also advisable to arrange for two or three parallel strips between each cell to make redundant contacts. This way, if one solder joint fails, the array will not stop working, as could happen with single jumper strips. An example of a cell where three jumper strips are possible is shown in Figure 4.4. In this design, three wide bars connect all the fine fingers allowing multiple paths for current to flow around a gap or defective finger. It is recommended that thin, preferably crinkled, strips of copper be sweat-soldered the entire length

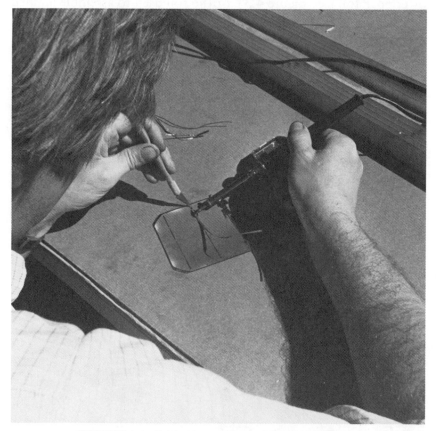

FIGURE 4.4—Soldering copper strips to a solar cell
(David Ross Stevens photo).

of the wide bands. These strips should also run at least halfway across the back of the next cell. This will allow a cell that accidentally gets cracked to continue producing power at nearly full output.

Instead of copper foil, it is possible to use small-diameter copper wire as the interconnections. B&G No. 24 to No. 26 wire is the most useful size. At least two pieces of wire and preferably three or more should be used. One source for this thin-diameter wire is multistranded No. 16 or No. 18 wire. Once the insulation is stripped, you have 20 or more strands of wire, prepared to the same length in one operation. The wire interconnections should be arranged so that they have a slight bend when the cells are positioned properly to allow for thermal expansion and contraction.

SOLDERING SOLAR CELLS

Most commercially available solar cells are designed to have contacts and wires soldered to them. The back contacts and front fingers are applied to the cells in a number of ways, including electrodeless nickel plating, vacuum metallizing and silkscreening. In addition, many manufacturers tin, or coat with solder, the contact points where jumpers are expected to be fastened. Ordinary electronic-grade (60% tin/40% lead) solder can be used in most cases, but a special solder which contains 2% silver in addition to 60% tin and 38% lead is the best choice if the cell contacts are silver. This will keep the hot solder from dissolving and possibly weakening the bond of the very thin silver fingers. Rosin core solder and rosin-based soldering flux can be used but can be difficult to clean off the cell faces once the array is soldered together. A water-soluble flux is available (Kester Solder Co., Chicago, Illinois) that is normally used by solar cell manufacturers. This flux simplifies cleaning the cells after the interconnections have been soldered.

A small pencil-type soldering iron of about 35 to 40 watts capacity is the most useful for this kind of soldering. The cone-shaped (pencil) tip or the small screwdriver-shaped tips seem best able to deliver a precise amount of heat to a small area.

The actual soldering of the copper foil pieces to the cells is not too difficult but requires fast work with a hot iron. As with any soldering job, it is most important that the surfaces are clean. It is best to clean solar cell surfaces with a cotton swab and isopropanol (rubbing alcohol) and let the surface dry before soldering. If the cell contact is already tinned, put a thin layer of soldering paste on the copper foil and sweat-solder the foil with the soldering iron. Figure 4.4 shows this operation. Use firm pressure, but not enough to crack the cell. No solder is necessary. If the cell contact is not tinned, tin the foil strip first. The strip must be tinned on every surface that will be soldered, but try to leave an untinned place where the foil jumps from one cell to the next, in order to increase the flexibility. It is best to solder all the foil tabs to the finger side of each cell first, then arrange the cells face down in the position desired for the array. Taping the cells face down to a piece of hardboard in the desired pattern will make it easy to align the cells properly. Then solder down the tab to the back of each cell. If small pieces of masking tape are used, they can be removed easily without danger of breaking the cells. It is best to make the contacts to the final cells in the array with the same kind of foil and then solder wires to the foil a centimeter or so away from the cell to reduce the strain.

If wires are used as the jumpers, they are soldered in a manner similar to that used for the foil strips, except that it is generally not necessary to pre-tin the wires before fastening them to the cells.

BUILDING A FLAT PLATE MODULE

One simple way to offer protection for the cells is to place them in a flat acrylic box. The collector box has a removable lid that is fastened down with screws and sealed with silicone grease. While not sealed tightly enough to withstand constant exposure to the salt spray on the deck of an ocean-going yacht, an array installed in such a box has been operating outdoors for two years without any decrease in electrical output.

Laying Out the Case

To construct the flat plastic module, the first step is to lay out the arrangement desired for the solar cells. For a 32-cell array, I usually use 4 rows of 8 cells each, but other arrangements are possible: 3 rows of 11 cells for 33 cells, 6 rows of 6 cells for a square 36-cell array, or 4 rows of 9 cells if a rectangular 36-cell array is desired. Other combinations are possible with unequal row lengths, *e.g.*, 3 rows of 6 plus 2 rows of 7 for an almost square 32-cell array. I prefer to put the cells in straight rows separated by thin acrylic strips. The strips keep the rows of cells separated without fastening the cells to the case, and they also help support the lid of the case so that it can't be pushed inward enough to press onto and possibly crack the cells.

Once an arrangement has been devised, you can calculate how big the case should be. The easiest way to do this is to actually fasten the cells into their final arrangement, lay them onto a flat surface, and measure the set, allowing about 1/2 inch between each row and another 1/2 inch at each edge. Figure 4.5 shows a case built for 3-inch circular cells. The dimensions given in the drawing will allow plenty of room for expansion and contraction. They also allow for some error in soldering the cells into absolutely

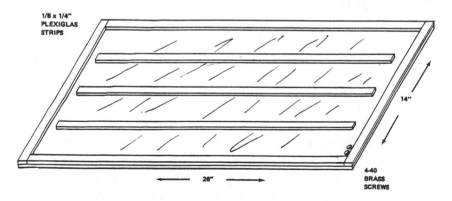

FIGURE 4.5—Drawing of the plastic case for a flat plate module. The dimensions shown are for 75-mm (3-inch) circular cells.

straight rows, but at the same time prevent the sets of cells from moving too far in relation to each other and the case.

Cutting the Plastic

Acrylic sheets (also called Plexiglas, trade name for Rohm and Haas plastic) come in a variety of thicknesses and qualities. The cast acrylic sheets are a little easier to use and glue, but are more expensive; the extruded sheets are cheaper and seem to be a bit stronger. The extruded sheets are 1/10 inch thick (called 100 mil by glaziers) and are commonly used to glaze storm doors where glass is too dangerous. This thickness is fine for solar cell arrays of the size discussed here. When using the cast acrylic, I generally use the 1/8 inch thickness, which is more than adequate for most purposes. If the array case will be subject to a great deal of abuse, 3/16 or even 1/4 inch could be used for the front and back covers, but you should use 1/8 inch for the spacer strips unless you plan to put some sort of backing material behind the cells to cushion them. For even more strength, polycarbonate sheets can be used instead of acrylic.

Cutting Plexiglas is quite easy. The easiest way to cut the 100-mil sheet is to score it with a utility knife and then break it along the score by bending it away from the scored side. Use a metal straight edge to get a straight scratch and score the sheet as deeply as possible. Unlike glass, acrylic can be scored by several passes over the same line, and if the knife cuts almost all the way through in spots, so much the better. Cutting the 1/4-inch-wide spacer strips by this method may be difficult unless you are proficient at working with the material.

Acrylic can also be cut with a hand saw or a saber saw, or even a table saw or portable circular saw. The problems here are chipping or cracking the plastic if too coarse a blade is used, or melting the plastic if too fine a blade is used. I have cut acrylic only to find the molten plastic and sawdust welded back together behind the saw blade. When using a circular saw, a veneer cutting blade seems to be the best compromise between melting and chipping.

Polycarbonate is very similar to acrylic when cut with a saw, but cannot be scored and broken as easily.

Gluing the Plastic Case

There are a number of commercially available glues for acrylics. Most of these work on a solvent principle, dissolving the acrylic slightly to make a sticky surface. I usually use a chlorinated hydrocarbon such as trichloroethylene or chloroform. Chloroform is probably the best solvent glue for acrylic, but it is extremely dangerous and is suspected of causing cancer. Avoid breathing the fumes or getting it on your skin. The commercial glues, while still very dangerous, do not require the extreme caution necessary in handling chloroform. The best solvent glue for polycarbonate is acetone, but methyl ethyl ketone will also work.

To use the solvent glues, place the pieces to be joined in the desired position and, using an eyedropper, let a few drops of the solvent run into the crack between the pieces. Capillary action will pull the solvent into place. Light but steady pressure on the joint will produce a complete bond. Do not move the pieces for 5 minutes, then examine the joint to see if it has been glued evenly. The glued parts will look wet and transparent when viewed through one of the shiny faces of the pieces, while the spots that are not actually in contact will reflect more light. You can apply a few more drops of solvent to these spots to glue them. It is possible, using this technique, to make a completely sealed, waterproof joint as strong as the plastic.

Fastening the Cover

When you have all the edge and divider strips glued onto the backing sheet, you are ready to install the solar cells and fasten down the cover. If you are sure that the array is soldered correctly and working properly, thus needing no maintenance for quite some time, you can attach the cover using the procedure just described. But a better idea is to fasten the top cover with screws.

After the plastic case is glued together, but before you install the cells, lay the top cover temporarily in place and drill and tap a set of holes around the entire outer edge of the case. I usually fasten the cover with 4-40 brass-headed screws. To ensure that all the holes will align, install two at opposite ends before drilling the rest of the holes. Drill and tap a couple of holes into the center divider strip to help hold the case together.

Installing the Solar Cells

Once the cover is prepared, it can be removed and the strips of cells laid into the case. Sheets of polyethylene foam packing material (1/16 inch thick) can be fitted into the bottom of the plastic case as a cushion behind the solar cells. (Do not use foam rubber for this purpose as it deteriorates in sunlight.) If you turn every other string end for end, the jumper wire from the back of the last cell in one string will be at the same end as the jumper wire from the front of the first cell in the next string, making the connections between strings short and easy to make. It is best to slip a piece of cardboard under the jumpers when soldering them in the case because acrylic cannot withstand much heat.

There are several methods for bringing the final wires out of the case. The easiest is to simply file a couple of notches into the top of one edge strip, one on each side of a cover screw, and lay the wires in these notches so that the cover clamps them in place. However, the danger with this arrangement is that someone might yank on the wires and break a cell where it is fastened to the wires. A neater solution is to attach a pair of binding posts to the back, e.g., a pair of 6-32 screws that are screwed from the inside through holes tapped through the rear cover. Use a little silicone caulk under each screw to seal the joint. I make the holes exactly 3/4 inch apart so a dual banana jack can be plugged onto the connection if desired. If it is small enough, you can attach a blocking diode inside the case to one of the binding posts. The cathode end of the diode (the end with the white band) goes to the positive (+) binding post while the anode end is fastened to the rear of the last cell in the string. Of course, the front of the first cell is connected to the negative (−) binding post.

After you have assembled and wired the array, try it out in the direct sunlight, or as bright a light as possible. Then clean the inside of the front cover and install it, using some silicone grease all aroung the edge as a sealant. Clear silicone caulk can also be used, as it doesn't adhere very well to the plastic and it is possible to pry up the cover if you have to get inside to repair the array.

BUILDING A HYBRID CONCENTRATOR ARRAY

Figure 4.6 shows a second type of array made at a Skyheat workshop. This hybrid array has two advantages over the flat array described above: (1) it can use a Winston concentrator to increase the amount of sunlight striking each cell, and (2) it has provisions for heating either water or air with the waste heat produced in the solar cells. We have made two slightly different versions of this collector: one with a sheet metal backing support and water tubes, and one with the cells mounted on a strip of wood. The second version can only be used for heating air, but it is simpler and cheaper to make. Both versions will be described here.

The Backing Support

The wooden backing support is simply a piece of wood about 1/2 inch wider than the cells and long enough to carry the desired string of cells. It is probably a good idea to limit the length of a single board to 4 or 5 feet, making the string of cells 16 cells long.

FIGURE 4.6–Skyheat workshop-fabricated hybrid solar cell array installed on a greenhouse (David Ross Stevens photo).

Two of these modules then make up one array. If the backing support is made exactly 4 feet long, it is possible to fix 16 three-inch cells onto the piece. Since each cell has a small flat on one end, the cells can be mounted slightly less than 3 inches apart, leaving a millimeter or so between cells.

A backing support bent from sheet metal will conduct heat much better than wood. A couple of small tubes soldered to it serve to carry water to cool the cells and deliver the heat to a hot water heater. Figure 4.7 is a drawing of one possible arrangement.

It is best to make the sheet metal support and the cooling tubes out of the same material to keep differential thermal expansion from warping the system, but small (1/4-inch-o.d.) copper tubing can be soldered into the corners of the backing plate and the combination doesn't warp. The dimensions shown in Figure 4.7 are for a concentrator system using 3-inch-diameter cells. Larger or smaller cells can be accommodated by changing the width, but the 1-inch depth of the legs is good for stiffness. 18-gauge galvanized sheet metal is thick enough for this use.

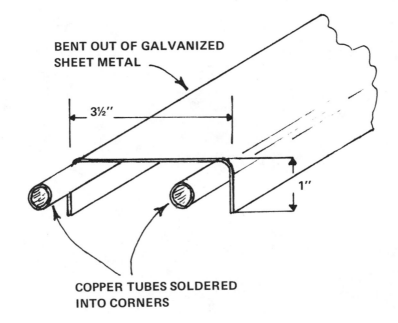

FIGURE 4.7—Sheet metal backing support.

Another possible backing support for this kind of concentrator array is an aluminum extension. Manufacturers of solar hot water collectors now make an extension 6 inches wide and flat on the top surface with a groove in the back that is designed to be fitted with a piece of 5/8-inch-o.d. copper tubing. This makes a rather simple and elegant holder for solar cells. Solar cell modules with this type of backing support have been successfully constructed at several workshops.

Isolating and Encapsulating the Cells

If a metal backing support is used for the solar cell array, the string of cells must not be allowed to touch the backing, as it would short the cells. Thus, the cells must be fastened to the backing in such a way that electrical insulation and adequate heat conduction are ensured. Silicone RTV (transparent silicone rubber caulk) is a very good electrical insulator, a moderately good conductor of heat, and has the added advantage of adhering extremely well to silicon cells. Based on the use of this material, we developed the following method to fasten and seal cells.

Step 1. Using a caulking gun, lay a bead of silicone RTV down the center of the backing plate, and quickly spread this layer to cover the entire plate about 1/16 inch thick (Figure 4.8). Speed is more important than perfection at this point.

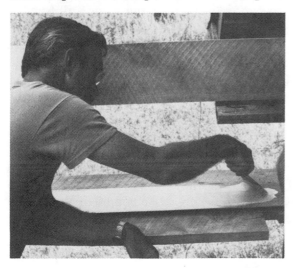

FIGURE 4.8—Spreading silicone RTV caulk onto the backing plate.

Step 2. Cut a strip of cloth about 1 inch wider and larger than the backing plate. (This strip can be prepared in advance so that no time is wasted while the silicone is setting up. Use polyester or some other synthetic cloth so any moisture that might get into the layer will not rot the fabric.) Press the cloth onto the fresh silicone layer (Figure 4.9).

Step 3. Coat the cloth with another silicone layer, taking more care this time to get a level surface (Figure 4.10). The purpose of these two steps is to create a barrier to keep the cells from accidentally touching the backing metal. For wood backs, Steps 2 and 3 can be eliminated if desired, although the cloth does strengthen and stabilize the silicone layer.

Step 4. Lay the solar cells carefully onto the wet silicone layer (Figure 4.11). It is best, if possible, to solder the copper jumper strips to the back of each cell and place the cells one at a time onto the backing. Then solder them together after testing each cell for proper operation and shorts to the metal base. Press each cell *very* gently into the silicone to work the air bubbles out from the back. (A few air bubbles are better than a cracked cell, however.) If you must solder the string together before embedding, have someone hold up one end of the string while you carefully place each cell in turn. It is possible, if you are quick enough, to remove a broken cell and replace it with a good one while the silicone is still tacky. It is almost impossible to remove a cell once the array is finished.

Step 5. When you are completely satisfied that the array is working properly, apply a top layer of silicone RTV over the entire array. I usually finish by rolling out a Mylar top film while the silicone is still fresh. (3M sells rolls of Mylar plastic as an accessory to their overhead projectors.) Simply place a bead of caulk on the cells and slowly roll the Mylar over it, in rolling pin fashion, to spread the silicone (Figure 4.12). With some practice, this system will produce a thin, uniform top surface without air bubbles trapped on the top of the cells. Make sure that the wires projecting from the two ends of the array do not put any strain on the connections to the end solar cells. I usually seal a couple of S curves in these wires so they are held in place by the silicone caulk.

If it ever becomes necessary to repair a cell connection, it is possible to cut away the top encapsulant in the area to be repaired (a razor blade works well). Once the repair is made, a dab of silicone

FIGURE 4.9–Placing the cloth onto the fresh silicone.
(Skyheat Workshop photo)

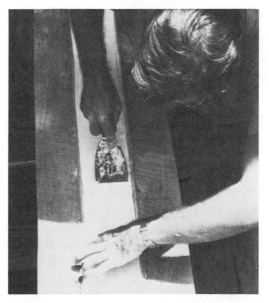

FIGURE 4.10–Smoothing out a new layer of silicone over the cloth,
completely embedding the cloth (Skyheat Workshop photo).

FIGURE 4.11—Laying the string of solar cells onto the freshly prepared backing (Skyheat Workshop photo).

RTV smeared over the spot will reseal the system. If you are not sure where the defective spot is, you can use a multimeter with one probe replaced with a needle to poke carefully through the encapsulant to touch the jumper strips. Connect the other probe to one of the wires coming out of the array, and probe each jumper in turn. (Carefully confine this poking to the copper jumper strips as you can break a solar cell by hitting it too hard with the needle.) The bad jumper will be either the one which shows no reading, or the one next to it, depending on which end of the jumper has the bad connection. I prefer to measure the short-circuit current with a meter when doing this kind of troubleshooting, as a bad solder joint may be leaky enough to allow the correct open-circuit voltage to be read, even though it will let virtually no current pass.

FIGURE 4.12—Coating the top of the cells with silicone and Mylar film. The Mylar film squeezes out a bead of the silicone RTV, eliminating voids and air bubbles (Courtesy Home Energy Workshop, Fort Collins, Colorado).

FIGURE 4.13—The finished product is a completely encapsulated set of solar cells (Courtesy Home Energy Workshop, Fort Collins, Colorado).

The Reflector Wings

Design

Once the long arrays are assembled and sealed, they are ready to use just as they are, but much more effective use can be made of these expensive solar cells if reflectors are added. For low concentration ratios, the design is not critical. Figure 4.14 shows the dimensions of several reflectors that can be used with this type of array. The first one is simply a pair of tilted flat reflectors that will almost double the amount of sunlight falling on the cells. This simple system, while very easy to build, doesn't have the acceptance angle of the second design; thus, the array tilt would have to be changed about four times a year for most efficient operation. The second system also has a concentration ratio of 2, but the curved surfaces are more effective in reflecting off-axis light onto the cells; thus the tilt angle need only be changed twice a year.

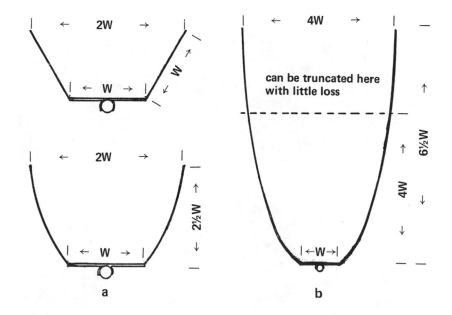

FIGURE 4.14–Dimensions of several reflectors: (a) 2-to-1 concentration ratio; (b) 4-to-1 concentration ratios.

This design works with reasonable efficiency even if set at a simple fixed angle (equal to the latitude from the vertical). The third system is simply a 4-to-1 concentration ratio version of the same Winston design. In this system, the reflectors are quite tall and will need some bracing to keep their shape. Also, the exact shape is more critical. Wood-backed arrays are not recommended for use with this high a concentration ratio, since the power input to the cells is quite high and the stagnation temperatures reached by the wood on a sunny day will decompose the wood and could actually cause a fire. In fact, wood-backed collectors could be dangerous in any hybrid system where the chamber above the wood is sealed off from the outside air.

Making the Reflector Wings

The easiest material from which to form the reflector wings is aluminum flashing. It is easily obtained at any lumber yard, stiff enough to hold its shape, and reasonably shiny. However, aluminized Mylar is much more reflective. The extra time and trouble taken to cover the reflectors with this material will make them considerably more effective, producing as much as 30% more electricity than a system with plain aluminum reflectors. Some brands of aluminized plastic are designed for exterior use and will keep their high reflectance for years.

When bending the aluminum reflectors to shape, first put a sharp kink where the reflector is fastened to the backing strip. You can improvise a long bending brake for this light-gauge material by clamping the aluminum between two boards with C-clamps and bending the protruding edge by hand or with a third board. The gentle parabolic curve in the metal can be approximated by bending the sheet around a round object. (We use a piece of 3-inch plastic sewer pipe.) The greatest bend is closest to the crease, and the springiness of the aluminum will keep the outer edge to a barely discernible curve.

The wings can be fastened to the array backing with sheet metal screws. (Be careful not to drill through a solar cell.) Then make the final adjustments to the wing shape and angle. Prop up the array so that you can look into it from about 15 feet away. You will see a distorted reflection of the solar cells in each wing if you look at the module from straight on-axis. Bend and adjust the reflectors until the blue solar cell images seem to fill the reflectors completely. By moving your head to one side, you can get an idea of how much off-axis sun angle the reflector system will accept.

If two or more of these modules are installed side by side, they can be mounted so the top edges of the reflectors just touch. The reflectors can then be fastened together to support each other, producing a very rigid set-up.

A Hybrid Hot Air System

Figure 4.15 shows two ways of ducting air through the collector to make a hybrid hot air system. The first incorporates a transparent cover over the entire array to create a series of plenum chambers out of the spaces between the reflector wings. The cover isn't a bad idea in any case as it will further protect the cells from the weather and keep the reflectors clean, at a slight loss in efficiency due to the reflection losses in the cover material. If the air is blown through this chamber, the cover should be double-glazed. In the second ducting method, the air is static above the cells and blown through the space behind them. Single glazing is sufficient if this method is used. The basic system is amendable to many variations, and it will be interesting to see what clever innovations develop.

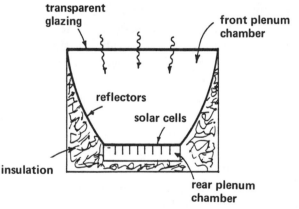

FIGURE 4.15—Two ways to duct air through a hybrid system, through either the front or rear plenum chamber.

Chapter 5
How Solar Cells Work

In order to understand how solar cells work, you first have to know something about the nature of sunlight. All light, including sunlight, is a form of electromagnetic radiation similar to radio waves or microwaves. The sun gives off this radiation simply because it is hot. This **black-body radiation** is composed of a broad mixture of different wavelengths, some of which can be seen by the naked eye (the visible spectrum) and many wavelengths shorter or longer than these. The sun is a black body—if it were cold, it would appear black, for it would only absorb radiation.

Figure 5.1 is a diagram of the solar spectrum. The curve labeled **AMO** shows the **air mass 0** spectrum the sun gives off as it appears in outer space, and the other curve is the **air mass 1 (AM1)** spectrum as seen from the surface of the earth. The difference is caused by our atmosphere. In this diagram, the visible spectrum is shown in the middle, from 400 to 700 nanometers in wavelength [a nanometer (nm) is 10^{-9} meters or one-millionth of a millimeter]. The red end of the visible spectrum is at 700 nm, while the shorter wavelength end at 400 nm appears blue or violet, with the rest of the rainbow colors between. Our eyes are most sensitive to wavelengths of around 500 nm where the peak of the solar spectrum on earth appears. Over half the energy that reaches the earth's surface from the sun is in the form of visible radiation.

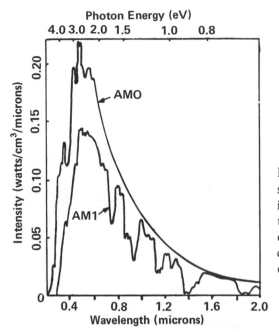

FIGURE 5.1—The solar spectrum. The dips in the intensity of the AM1 spectrum seen from the surface of the earth are mostly caused by water vapor in our atmosphere.

Wavelengths shorter than 400 nm are called **ultraviolet (UV)** and are present in the solar spectrum in small but significant amounts. Ozone and most transparent material absorb or filter out most of these energetic wavelengths—fortunately, since organic material and living things can be damaged or destroyed by short-wavelength ultraviolet light. The small amount of UV light that does make it to the earth's surface is responsible for the tan or sunburn we receive when we are exposed to the sun.

Wavelengths longer than 700 nm are called **infrared** and, while invisible, can be perceived by our skin as radiant heat. As the diagram shows, a great deal of the total energy from the sun is infrared radiation. Large bands of infrared are absorbed by water vapor, carbon dioxide and other substances in our atmosphere, but, because most of this absorption takes place at longer wavelengths, a solar device that doesn't effectively collect wavelengths longer than 2 microns (2000 nm) suffers only a small loss in efficiency.

According to the theory of quantum mechanics, light can be looked upon as being composed of particles called photons. Each photon (which travels through a vacuum at the speed of light) has no mass but is a packet of energy related to the wavelength of the light. The shorter the wavelength, the larger the packet. The energy of individual photons of different wavelengths is shown at the top of Figure 5.1. The energy is expressed in electron volts. (One electron volt is the energy an electron acquires when it accelerates in a vacuum across a potential of 1 volt. This unit of energy is commonly used by solar cell physicists since it is a convenient size when considering individual electrons.)

SOLID-STATE PHYSICS

Although the solar cell was accidentally discovered in the nineteenth century, and inefficient versions of selenium and cuprous oxide photovoltaic cells were investigated and even used commercially in the early part of this century, it was only after the development of modern solid-state theory and the **band model** of semiconductors that the inner workings of photovoltaic cells were understood.

The band model of solids, a simplified version of which will be given here, is based on the idea of a crystal with all the atoms in a fixed regular pattern. The individual atomic nuclei can vibrate around a fixed spot in this three-dimensional lattice pattern, but normally cannot completely jump out of place. The powerful electrostatic forces tie most of the negative electrons in a particular atom very closely to the positively charged nucleus. However, the outermost electrons (called the valence electrons) can be looked upon as "delocalized," *i.e.*, not belonging to any particular atom, but to the crystal as a whole.

These valence electrons balance the remaining positive charges of all the nuclei. It takes a certain amount of energy to remove one of these electrons from the crystal because of the electrostatic attraction. This energy can be represented on a diagram, such as Figure 5.2, as a vertical distance. In this figure, which represents the band diagram for a metal, zero energy is taken as the energy of an electron at rest in empty space and is called the **vacuum zero**. The energies of the electrons in the crystal are all below this zero,

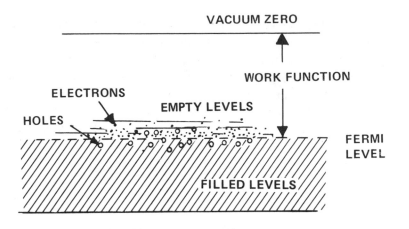

FIGURE 5.2—Energy level diagram for a metal.

and the kinetic energy of a fast-moving electron in vacuum would be above the zero.

One of the conclusions of quantum mechanics is that electrons (or anything else for that matter) cannot have a continuum of energies but are allowed only certain selected energies (these are called **energy levels** in the diagram). Another quantum mechanical principle is that no two electrons can occupy the same energy level.

It is possible to calculate the number and positions of the energy levels of a particular metal, and, if a count is made, it turns out that in a metal each atom contributes more energy levels than electrons. Such a situation is diagrammed in Figure 5.2. Since the band of valence levels is greater than the number of electrons, some of the levels must be empty. The electrons would preferentially occupy the lowest lying levels and, at absolute zero when none of them are moving, would fill up the band to the **fermi level**. The fermi level is defined as the level where the probability of finding an electron is one-half. In order to move, the electrons must have a bit more energy (their kinetic energy) and would occupy a level just above the fermi level. Since there are a great number of empty levels at this energy, the electron has a large number of possible kinetic energies and can move very freely throughout the crystal. These moving electrons can produce an electric current, so a metal would be expected to be a good conductor of electricity, capable of carrying a large current.

Each electron from below the fermi level that jumps to a higher, empty level, leaves behind an empty level. These empty levels in the "sea" of electrons are called **holes**. A hole is a level that an electron could occupy but currently does not. A hole "moves" when an electron falls into it, creating a new hole. So a hole can be looked upon as a particle similar to an electron with a positive charge, a mobility, and even an effective mass. Holes can carry an electric current and in some metals are the dominant carriers.

If the number of available levels is exactly the same as the number of electrons for each atom, the situation depicted in Figure 5.3 results. This figure also shows a band of completely empty energy levels at some higher energy. Since the number of excited electronic states in any atom is theoretically infinite, there will always be such a **conduction band**, sometimes separated from the **valence band** by a **band gap**. The valence band is completely filled so there are no nearby energy levels for an electron to occupy. The electrons cannot move and cannot create any holes. Therefore no conduction can take place, and the material is called an insulator. The energy of the band gap, like all the energies shown in the diagrams, is in electron volts.

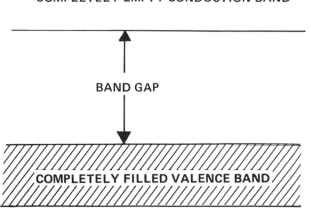

VACUUM ZERO

COMPLETELY EMPTY CONDUCTION BAND

BAND GAP

COMPLETELY FILLED VALENCE BAND

FIGURE 5.3—Energy level diagram for a perfect insulation.

If the band gap is sufficiently narrow (below 2.5 electron volts or so), it is possible for a particularly energetic electron to jump from the top of the valence band to one of the empty levels in the conduction band (see Figure 5.4). The electron could be thermally excited by the vibrational movement of the atoms in the crystal lattice, or a photon of light of sufficient energy could be absorbed by the crystal and cause the electron to jump. Once the electron is in the upper conduction band, it is free to move and can act as a carrier of electricity. In addition, the hole left behind in the valence band can also be a current carrier. A material having these properties is called a semiconductor.

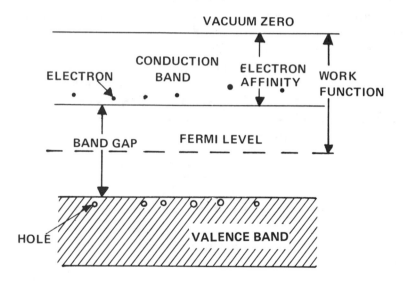

FIGURE 5.4—Energy band diagram for an intrinsic semiconductor.

The fermi level shown in the figure is in the middle of the band gap. Above the fermi level, fewer of the levels are occupied, while below it, most are. The unoccupied levels are the holes.

As the temperature of a solid increases, the most energetic electrons and holes can be found further from the fermi level, but the position of the fermi level remains fixed. In the case shown in Figure 5.4, there are no energy levels at all near the fermi level, but if there were, there would be about a 50% chance that they would be occupied. Only the most energetic electrons and holes are actually occupying levels. [The most important point to remember is that the closer a band is to the fermi level, the more conduction electrons (or holes, if it is below the fermi level) that will be found in the band.] For a more complete mathematical discussion of these solid-state concepts, see Kittel's *Introduction to Solid-State Physics*.

DOPING

The electrical conductivity discussed in the last section is called **intrinsic conductivity** because it is an intrinsic property of the particular material and its crystal lattice. An ideal **intrinsic semiconductor** cannot exist in reality, since all materials have some impurities, no matter how carefully prepared. The impurity atoms will occupy various spots in the crystal lattice and disturb the perfection of the lattice. In addition, these foreign atoms will usually contribute a different number of levels and/or electrons to the system than the normal atoms. For example, in a single crystal of silicon, each silicon atom contributes exactly 4 valence levels, 4 conduction levels, and 4 electrons since the valence of silicon is 4.

If an atom of boron (which has a valence of 3) is added to the crystal, it will contribute only 3 electrons to the system. It will also contribute 3 valence levels, plus 1 level which will be created in the band gap slightly above the top of the valence band. This situation is diagrammed in Figure 5.5a. These extra levels are capable of accepting an electron from the valence band and are called **acceptor levels**. Since the acceptor level is associated with a particular impurity atom, the electron occupying it is trapped and cannot move. However, the hole created when the electron leaves the valence band can move and carry an electric current. Silicon doped with boron is called a **p-type semiconductor** since positive holes carry the current. This process of deliberately adding a known impurity to a pure semiconductor is called **doping** and the resulting material is called an **extrinsic semiconductor**. Similarly,

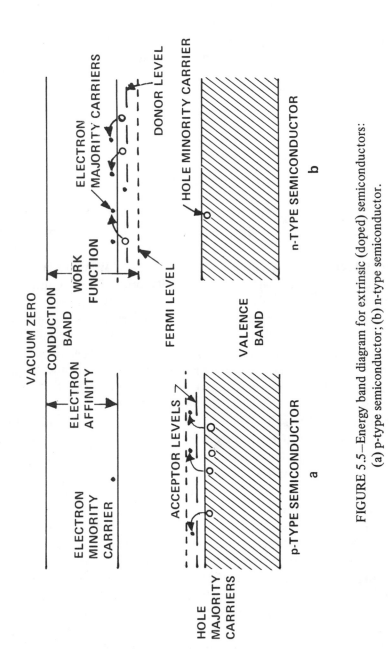

FIGURE 5.5—Energy band diagram for extrinsic (doped) semiconductors: (a) p-type semiconductor; (b) n-type semiconductor.

if a small amount of phosphorus is added to the silicon crystal, the atoms of phosphorus (which have 5 valence electrons) will contribute 4 electrons and 4 levels to the valence band, and the extra electron will occupy a level near the bottom of the conduction band as shown in Figure 5.5b. It takes very little energy for the extra electron to jump to the conduction band from this **donor level**, so-called because it can donate conduction electrons to the system. This type of material is called an **n-type semiconductor** since negative electrons carry the electric current.

Notice that Figure 5.5a shows p-type material, the fermi level is shown near the valence band, and the vast majority of current carriers are holes. In Figure 5.5b which shows n-type material, the fermi level is near the bottom of the conduction band, and the majority of the carriers are electrons. In p-type semiconductors, the positive holes are the **majority carriers** and the electrons are the **minority carriers**. In n-type semiconductors the negative electrons are the majority carriers and the holes are naturally called minority carriers.

Finally, notice on the diagram, the quantities marked **work function** and **electron affinity**. The work function is defined as the energy difference between the fermi level and vacuum zero, and the electron affinity is the energy difference between the bottom of the conduction band and vacuum zero. These definitions are true for all semiconductors. For metals, the work function and electron affinity have the same value.

JUNCTION PHOTOVOLTAIC CELLS

Figure 5.6 shows the construction of a typical n-on-p junction solar cell. This is the most common type of cell and typically would be made by taking a thin slice or wafer of p-type single-crystal silicon and diffusing phosphorus atoms into the top surface. This can be done by heating a batch of the wafers in a **diffusion furnace** in the presence of phosphorus-containing gas. The back contact and the top fingers shown are used to connect the solar cell to the external circuit. A more detailed description of the construction of these cells is given in the next chapter.

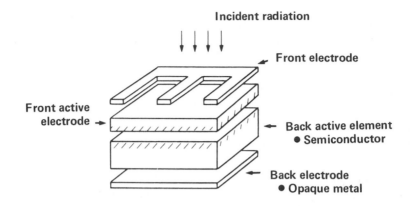

FIGURE 5.6–Cross section of a solar cell showing its construction.

The energy band diagram for this n-on-p cell (called a **homo-junction solar cell**) is shown in Figure 5.7. This diagram is made by taking the diagram for the n-type semiconductor and putting it together with the one for the p-type material in such a way that the fermi levels line up. Notice that when this is done, there is a difference in the vacuum zero levels for the two materials. This represents the **built-in potential** that now exists between the materials because of the difference in the work functions of the two types of semiconductors. Between the n and p layers is a **depletion layer** where there are essentially no carriers–neither electrons nor holes. This layer, where the bands are bent, is the actual junction. The front and back metal contacts must be **ohmic contacts**: that is, contacts that do not impede the flow of electrons into or out of the semiconductor. In the dark, if the junction device is connected to a battery so the front fingers are made positive and the back negative, the situation shown in Figure 5.8a would result, where the applied potential is added to the built-in potential and the uphill barrier for the electrons is increased, resulting in no current flowing except for the tiny current of electrons thermally excited from the valence to the conduction band. At room temperature this current would be a fraction of a microamp. On the other hand, if the battery is reversed so that the front is negative and the back positive, the applied potential can

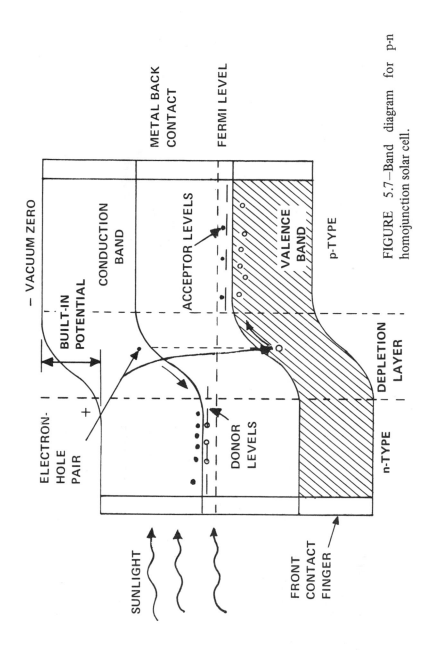

FIGURE 5.7—Band diagram for p-n homojunction solar cell.

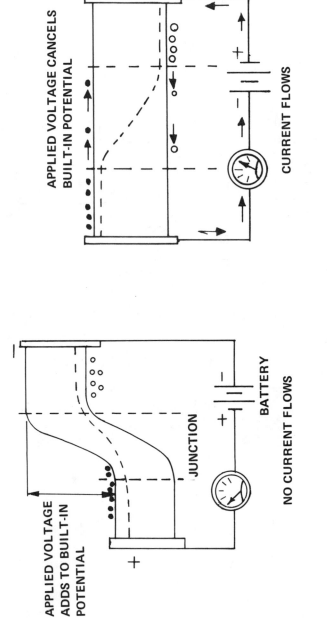

FIGURE 5.8b—Forward bias rectifier.

FIGURE 5.8a—Reverse bias rectifier.

be made large enough to cancel out the built-in potential and, as Figure 5.8b shows, there is no longer any barrier to the flow of either holes or electrons and the device becomes a good conductor of electricity. So a solar cell is a **junction diode** or **rectifier** letting electricity pass in only one direction in the dark. The dark curve in Figure 5.9 is a typical current voltage plot for this type of junction. The current really starts to increase in the forward direction when the applied voltage gets close to the built-in potential. For silicon junction diodes, this is around 0.6 volt.

If light falls on the solar cell, those photons with energy greater than the band gap width can excite an electron from the valence to the conduction band and create a hole–electron pair. The conduction electron, if it is created in the depletion layer, will slide "downhill" toward the front n layer, and the hole will (like an air bubble in water) slide "uphill" to the p layer, so the charge carriers will be separated and a current will flow. The ohmic front contact allows the electron to flow through an external circuit to the back contact where it can recombine with a hole, leaving everything the way it started. Even if the conduction electron is created away from the junction, if it happens to drift toward the junction, it will be carried down as before and produce a current in the external circuit.

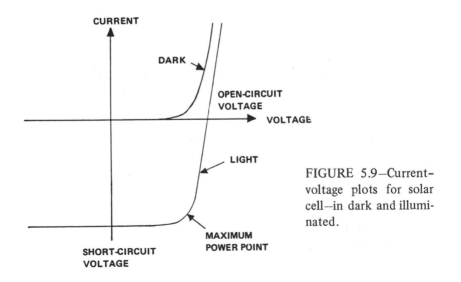

FIGURE 5.9—Current-voltage plots for solar cell—in dark and illuminated.

If the external circuit is simply a wire and has no appreciable resistance, the current that flows is called the **short-circuit current (I_{SC})** and is directly related to the number of photons of light being absorbed by the cell. This is why the short-circuit current is directly proportional to the light intensity.

If there is no external circuit, the incoming photons will still create hole–electron pairs and they will still travel downhill to the p and n layers as before, but they will pile up there since there is no external wire to complete the circuit. The extra negative electrons in the n layer and the extra positive holes in the p layer will cause a situation analogous to that shown in Figure 5.8b, except in this case the battery is the solar cell itself. Soon the voltage difference between the front and the back of the cell will become large enough to flatten the barrier sufficiently to allow some of the holes and electrons to leak back. When the number of carriers leaking back is equal to the number being generated by the incoming light, an equilibrium voltage will be reached. This voltage is called the **open-circuit voltage (V_{OC})** and is shown on the light current–voltage curve on Figure 5.9. The open-circuit voltage varies only slightly with large changes in light intensity.

If the resistance of the external circuit can be varied from zero (short circuit) to extremely high (open circuit), the entire current–voltage curve can be plotted. The power given out by the solar cell is defined as the voltage times the current:

$$P_{(watts)} = V_{(volts)} \times I_{(amps)}$$

For the short circuit, there is no voltage and hence no power output from the cell. Similarly, for the open circuit, there is maximum voltage but no current, so again there is no power output. Somewhere between lies the best combination of voltage and current to produce the maximum power. That point is shown at the bend in Figure 5.9. This maximum power output occurs when the resistance of the external circuit equals the internal resistance of the cell.

One way to judge the quality of a solar cell is by measuring the **fill factor (ff)**. The fill factor is defined as the actual maximum power divided by the hypothetical "power" obtained by multiplying the open-circuit voltage (V_{OC}) by the short-circuit current (I_{SC})

$$ff = \frac{P_{max}}{V_{OC} \times I_{SC}}$$

For a good silicon solar cell, the fill factor will normally be above 0.75.

FACTORS WHICH INFLUENCE EFFICIENCY

Band Gap Width

One of the most important factors influencing the theoretical efficiency is the band gap width of the semiconductor material. Remember that only photons of energy greater than the band gap can produce current carriers. This means that only light of wavelengths shorter than some critical maximum can be used to produce a photocurrent. In fact, most semiconductors are transparent to light redder than this cutoff wavelength, for the photons will pass right through the material without interaction. Therefore, small band gap materials will absorb more of the red end of the solar spectrum and would appear to be more efficient users of sunlight. However, the open-circuit voltage of a cell is limited by the band gap width of the cell material. Also, the energetic photons of short wavelength cannot produce any more current than the photons barely able to raise an electron to the conduction band, thus they are used less efficiently by narrow band gap materials (the excess energy of the photon shows up as heat, heating the cells). The best cell will have a compromise band gap width.

Dr. Dan Trivich of Wayne State University, in 1953, made the first theoretical calculations of the efficiencies of various materials of different band gap widths based on the spectrum of our sun. Figure 5.10 is an updated version of those calculations. As the diagram shows, silicon has too narrow a band gap for optimum efficiency, but 28 years of development have pushed the actual working efficiency close to the theoretical.

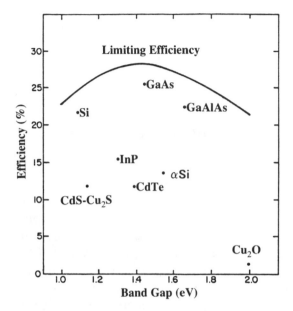

FIGURE 5.10- Efficiencies versus band gap for various kinds of semiconductor solar cell materials. Dots are one-sun measured values.

Recombination

A major factor which influences the actual attainable efficiency is recombination. Not all the hole–electron pairs created make it all the way to the metal contacts; some of the carriers will recombine with a carrier of opposite sign, eliminating both. This will usually happen at a defect or impurity in the crystal called a **recombination center**. It is particularly common to find these centers at the junction or at the surface of the material, but recombination can happen even in the bulk of the material. These mechanisms are called, respectively, **junction**, **surface** or **bulk recombination**. It is to minimize this recombination that high-quality single-crystal wafers are used in making solar cells.

The design of the finger pattern is a compromise between one that is very open, so as not to block out much light, and one of very low resistance, so that little of the power output of the cell is wasted on resistance heating of the fingers.

Reflectivity

Most semiconductor materials have a high reflectivity of the light they should be absorbing. Most manufacturers etch the surface of the cell to roughen or "texturize" it and apply antireflection coatings to decrease reflection losses. These coatings, similar to the coatings on good camera lenses, are usually silicon monoxide or silicon nitride for silicon cells.

HETEROJUNCTIONS

The n and p layers in a solar cell do not have to be made of the same material, but can be composed of two different semiconductors. Such a device is called a **heterojunction**. Figure 5.11 shows the band diagram for a heterojunction. The device operates in essentially the same way as a homojunction. The use of two different semiconductors gives a great deal of design flexibility; if

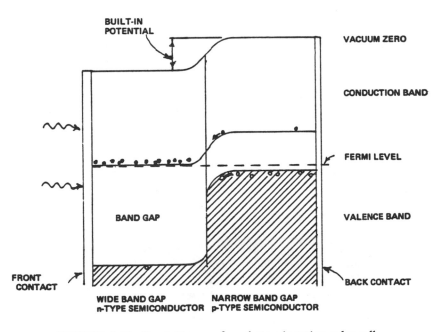

FIGURE 5.11—Band diagram for a heterojunction solar cell.

the top material has a wide band gap, it can serve as a window to let the light penetrate to the second layer. This layer is selected based on its efficiency as an absorber of sunlight. With the proper choice of materials, the device can have a larger built-in potential and, hence, a higher open-circuit voltage than a homojunction made from the narrow band gap semiconductor. Also, some semiconductors, such as cuprous oxide or tin oxide, are available in only one form, either p- or n-type, and cannot be made into homojunction cells. Heterojunctions can exploit the best properties of such materials. Chapter 7 discusses several new developments in heterojunction systems.

SCHOTTKY BARRIER JUNCTIONS

The last semiconductor device we will consider in detail is the **metal-to-semiconductor** or **Schottky barrier junction.** Figure 5.12 shows such a junction, which is formed when a low work function metal is placed on a p-type semiconductor. A high work function metal on an n-type semiconductor will also produce a Schottky barrier, which would have a band diagram resembling an upside-down version of the one shown. The important point here is the

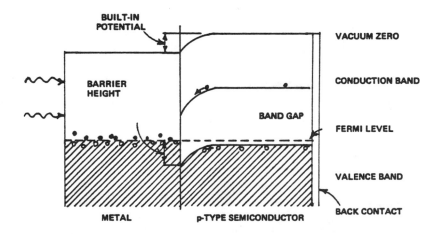

FIGURE 5.12—Band diagram for a Schottky barrier junction solar cell.

creation of a barrier to the flow of majority carriers from the semi-conductor to the metal and to the flow of all carriers from the metal to the semiconductor. Also, the band bending in the semi-conductor produces a built-in potential that determines the open-circuit voltage in the solar cell. The built-in potential is simply the difference in work function between the metal and the semicon-ductor while the barrier height is the energy difference between the work function of the metal and the top of the valence band (for p-type semiconductors) or the bottom of the conduction band (for n-type). The quality of the solar cell, as measured by the reverse dark current and the fill factor, is dependent in part on the barrier height. If a high work function metal is put onto a p-type semiconductor (or a low work function metal onto an n-type material) an **ohmic contact** with no barrier at all is the result.

The first solar cells—the cuprous oxide cell and the selenium cell—are both examples of Schottky barrier cells, but so far no efficient and inexpensive Schottky barrier solar cell has been con-structed.

ADVANCED SEMICONDUCTOR DEVICES

There are a number of other semiconductor junctions that have been considered for use in solar cells. Two promising devices are the metal-insulator-semiconductor (MIS) and semiconductor-insulator-semiconductor (SIS) junctions where a thin insulating layer is placed in the middle of the junction. Current carriers pass through this insulator by a quantum mechanical process called **tunneling**. Some scientists believe that all Schottky barrier junc-tions have an insulating layer and are MIS structures. If the insulat-ing layer is of the proper thickness, around 20 Å, the open-circuit voltage of a cell can be increased without any significant loss in short-circuit current.

RECOMMENDED READINGS

Angrist, Stanley W. "Photovoltaic Generators," Chapter 5, *Direct Energy Conversion*, 2nd edition (Boston: Allyn & Bacon Inc., 1971).

Chalmers, Bruce. "The Photovoltaic Generation of Electricity," *Scientific American* 235:34–43 (October 1976).

Hovel, Harold. "Solar Cells," in *Semiconductors and Semimetals, Volume II*, R. K. Willardson and A. C. Beer, Editors (New York: Academic Press, Inc., 1975).

Kittel, Charles. *Introduction to Solid-State Physics*, 3rd edition (New York: John Wiley & Sons, 1968).

Landsberg, P. T. "An Introduction to the Theory of Photovoltaic Cells," *Solid-State Electronics* 18:1043–1052 (1975).

Chapter 6

How Solar Cells are Made

The present processes for manufacturing solar cells are complicated, energy-consuming and expensive. This chapter describes in detail the current techniques for single silicon photovoltaic cell manufacture and examines the improvements now under development. The step-by-step procedure for making cells follows.

STARTING MATERIAL

You start with sand. Silicon, the second most abundant element in the earth's crust, is present in almost all rocks and minerals, but the most convenient starting material is silicon dioxide in the form of white quartzite sand. If the sand is pure enough, the subsequent purification steps are less complicated.

Metallurgical-Grade Silicon

Sand is reduced to silicon in an electric arc furnace. A carbon arc reacts with the oxygen in the silicon dioxide to form carbon dioxide and molten silicon. This common industrial process produces a metallurgical-grade silicon, with about 1% impurities, that has a number of uses in the steel and other industries. Though relatively inexpensive, this silicon contains far too many impurities for use in the electronics industry or to make solar cells. Many impurities are the result of contamination of the sand or are introduced during processing and could be eliminated by quality control and careful choice of raw materials.

Semiconductor-Grade Silicon

The electronic properties of silicon semiconductor devices are determined by a small percentage of **dopant** atoms in the crystal lattice. In order for the devices to be built, the number of impurities in the silicon starting material must be small compared to the dopants to be added. This means that semiconductor-grade silicon must be hyperpure; the residual impurities are measured in parts per billion. The most common way to produce silicon of this quality is by the thermal decomposition of silane or some other gaseous silicon compound. A seed rod of ultrapure silicon is heated red-hot in a sealed chamber and the purified silicon compound is admitted. When the molecules of this compound strike the rod, they break down to form elemental silicon which builds up on the rod. When the rod has grown to the desired thickness, it is removed, ready to be made into a single crystal.

An alternative process for purifying silicon is **zone refining**. In this process, a rod of metallurgical-grade silicon is clamped in a machine with a moving induction heater coil. The coil melts a zone through the entire rod near the bottom end and then slowly travels upward. The silicon melts at the top edge of the zone and solidifies at the bottom edge. When the liquid solidifies, impurities in the melt are excluded from the new crystal lattice, so the molten zone picks up the impurities and sweeps them to the top of the rod, which can be discarded.

Both methods of purifying silicon are expensive and energy-consuming; the resulting semiconductor-grade silicon presently costs $80 per kilogram. Solar cells currently are made from this type of silicon, which is becoming in short supply because of the growth of the semiconductor industry. Chapter 7 discusses several ways to make solar cells from lesser-grade materials.

FIGURE 6.1—Semiconductor-grade silicon. This highly refined material is purer than fine gold and is becoming more expensive as supplies fail to keep up with demand.   (Courtesy ARCO Solar, Woodland Hills, California)

GROWING SINGLE CRYSTALS

The boundaries between crystals in silicon act as traps for the electrical current carriers; so, for best performance, all semiconductor devices and most currently available solar cells are made from single crystals of silicon. Over the years, the techniques for growing silicon crystals have been developed to the point where crystals six inches in diameter and four feet long are routinely grown by the **Czochralski process.** In this method, a seed crystal is dipped into a crucible of molten silicon and slowly withdrawn, pulling a large round crystal as the molten material solidifies on the bottom of the seed. Feedback controls adjust the pulling speed and the temperature of the melt to produce a crystal of a given size. The total mass of the crystal is determined by the size of the pulling machine and the amount of material the crucible can hold. A small amount of dopant impurity is usually added to the silicon during this step to produce the desired electronic properties in the single crystal.

FIGURE 6.2—A single crystal of pure silicon being pulled from a crucible of molten material. Virtually all solar cells commercially produced use this Czochralski process.   (Courtesy ARCO Solar, Woodland Hills, California)

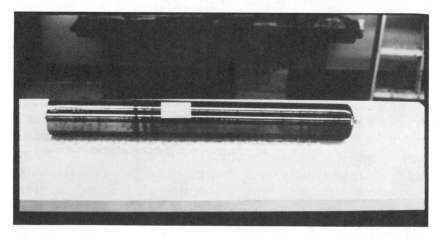

FIGURE 6.3—An ingot produced by the Czochralski process. The standard diameter of these ingots has increased over the years from 2¼ and 3 inches to 4¼ and even 6 inches as the technology has improved. (Courtesy ARCO Solar, Woodland Hills, California)

SLICING THE CRYSTAL INTO WAFERS

Because most semiconductor devices, including photovoltaic cells, must be extremely thin, the silicon crystal must be sliced into thin wafers. The thickness of the wafer—on the order of 400 microns (about 0.016 inch)—is determined chiefly by the fragility of the material. Thinner wafers would be difficult to handle. Slicing or "wafering" is most commonly performed on a machine with a single blade made of thin metal coated with diamonds. On most of these saws the cutting edge is the *inside* edge of a washer-shaped blade. Thus with these "i.d." saws, as they are called, the outer edge of the disc is rigidly supported in a heavy ring to reduce flexing and give a straight, smooth cut. A liquid lubricant is pumped continually into the saw cut to cool the materials and

FIGURE 6.4—Slicing a silicon single crystal by means of a diamond i.d. saw. This process is time-consuming and wasteful of material, the saw cuts are actually as wide as the wafers produced. (Courtesy ARCO Solar, Woodland Hills, California)

wash away the sawdust. The process, although automated, is time-consuming since one crystal will yield thousands of wafers, cut one at a time. It is also wasteful, as half the high-purity silicon is lost as sawdust. Techniques using multiple cutting blades or wires are in development; some will yield both thinner wafers and less waste.

POLISHING AND ETCHING THE WAFER

The surfaces of the solar cell must be specially prepared to ensure the proper electrical and optical properties. Originally, solar cells were highly polished using fine abrasives and chemical polishing etchants. A number of these mixtures, first developed for the microelectronics industry, leave a mirror-like surface that forms a good junction with the metals used for electrical contacts. Recently, texturized surfaces, which look like a mountain range under the high magnification of an electron microscope, have been found to be more effective absorbers of light. Some manufacturers, like Motorola, have dispensed with grinding and polishing completely, relying on a simple etching procedure to produce the desired surface. A number of other surface treatments, like ion milling, are being examined by manufacturers.

FORMING THE p–n JUNCTION

p-Type silicon is the starting material for most photovoltaic cells. During crystal growth, a small amount of boron is incorporated into the crystal lattice. To make a p–n homojunction cell, the top few microns of the wafer must be made n-type. This is normally achieved by incorporating more phosphorus atoms into the top layer of material than there are boron atoms. The excess phosphorus, being an electron donor, will make the layer n-type. To accomplish this, a rack of the wafers is placed in a **diffusion furnace** and heated to a high temperature in the presence of a phosphorus-containing gas. At this temperature, still well below the melting point of silicon, the individual atoms move and vibrate wildly within the crystal lattice, and foreign atoms that strike the

FIGURE 6.5—A diffusion furnace in which the p-n junction is formed. One such furnace can process over 1000 cells at a time. (Courtesy Motorola Semiconductor Group, Phoenix, Arizona)

surface diffuse slowly into the bulk of the material. If the temperature and time of the exposure to the gas are properly controlled, a uniform junction can be formed a known distance into the wafer. One furnace can form junctions in an entire rack of solar cells at one time. Of course, the surface that will later be the back of the cell must be protected to keep a junction from forming there also. Something as simple as sealing two wafers back to back can accomplish this. Sometimes a p+–p junction is also diffused into the back of the cell to create a back surface field and improve the collection efficiency for the light-generated electrons.

A newer method of forming the front junction is **ion implantation**. This method has been used for a number of years to make integrated circuits, but just recently has been applied to solar cells. The ion implanter, a machine similar to a small particle accelerator, shoots individual ions at the surface of the wafer. (In this case the ions are atoms with one electron missing.) The depth to which the ions penetrate the front of the silicon can be controlled by changing the speed at which the ions hit the surface, so the thickness and characteristics of the top layer of the cell can be carefully tailored. As automated ion implanters become more common, this junction formation technique will become an important one, especially for high-quality cells.

APPLYING THE FINGERS AND BACK CONTACT

Once the junction is formed, you have a working solar cell, but to make the cell useful, contacts must be placed on the wafer so the device can be connected to an external circuit. Both the front and back contacts must be **ohmic contacts** with as low a resistance to electric current as possible. The contact material must also adhere very well to the silicon and must withstand soldering (or spot-welding in some cases). The back contact can be a solid metal coating, since light does not have to pass through this area. In fact, a shiny metal contact will reflect back through the silicon any light that does not get absorbed in the first pass through the cell, allowing a thinner wafer of silicon to be used.

A great deal of work has been done in the design and fabrication of the contacts to solar cells. Each manufacturer has its own technique, but the goal of an inexpensive, reliable, rugged contact system has yet to be achieved. The most reliable finger contacts seem to be those made from palladium/silver layers that are vacuum-evaporated through a mask or photoresist to make the initial thin layer. Then, once the pattern is established, a thicker metal layer is electroplated on top. Fingers made in this way can be very narrow and close together, blocking out little light while still maintaining a small resistance.

Another contact system incorporates electrodeless nickel plating. The solar cell is covered with wax using a silkscreen so that the bare silicon is exposed in the areas where the fingers are desired. Then the cell is simply dunked in a bath containing nickel compounds. A chemical reaction deposits pure nickel on the uncovered silicon, producing the fingers. After the wax is removed, the fingers are tinned with solder. This system, while simple and relatively inexpensive, produces contacts that don't always adhere to the silicon as well as would be desired.

Other methods of applying the contacts include silkscreening a metal-containing paint and then baking the wafer to drive off the organic components, leaving the metal frit. A great deal of research is still under way to develop better contact systems.

THE ANTIREFLECTION COATING

Silicon is a very shiny material, having a grey metallic appearance. It reflects 35% or so of all the light that falls on it. This is lost light that could have generated electricity. So, presently, all solar cells are coated with an antireflection coating. This same antireflection coating has been used for decades to coat the lenses in good cameras, binoculars, and other high-precision optics. These coatings are transparent layers with a thickness of about one-fourth the wavelength of light. (A typical thickness is less than 100 nanometers, or 0.1 micron.) The coating could be made of any hard transparent material, but the preferred substances adhere well to silicon, have the right **refractive index** (they bend light properly), and are relatively easy to apply in accurately controlled layers.

Silicon monoxide, titanium dioxide and some other coatings can be applied to the silicon in a vacuum coating process such as **vacuum evaporation** or **sputtering**. In vacuum evaporation, a substance is heated to a high temperature in a vacuum system. The molecules of the substance "boil" off the surface and travel in straight lines until they encounter a cool surface to which they adhere and build up a thin layer. Thin coatings of metals and many ceramic substances can be made in this fashion. Sputtering is similar but uses a high voltage and a radio frequency field to knock molecules off a target material and deposit them on a material placed at the opposite electrode. This technique can be used to deposit refractive materials that cannot be evaporated in a vacuum system.

Another technique used for forming the antireflection coating is to make it out of the top layer of silicon. Silicon can react with oxygen- or nitrogen-containing gases to form silicon dioxide or silicon nitride. Although it is difficult to form a sufficiently thick layer of silicon dioxide by this process, silicon nitride antireflection coatings are used on commercial solar cells.

A final refinement used commonly in space-quality cells and by at least one manufacturer of terrestrial cells is the multilayer anti-reflection coating. Three or more layers of a set of two different transport substances can be used to cancel out the reflected light almost completely, giving a reflection loss of less than 3% compared to the 35% for the bare silicon.

ASSEMBLY INTO MODULES

At this point, the cell is completely fabricated and ready to be used. Most solar cells are assembled into modules by the individual cell manufacturers, and the design of the finger pattern of the cell is based on the particular module configuration of the company involved. Chapter 1 describes the characteristics of such modules in some detail.

In general, solar cell manufacturers are attempting to apply more and more automation to the module assembly process, which comprises some 40% of the cost of the finished product. Automatic

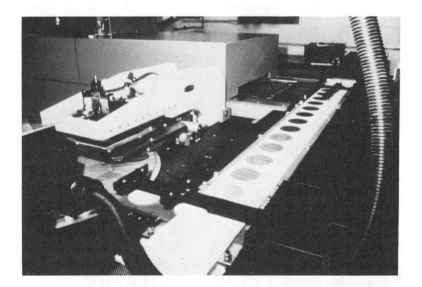

FIGURE 6.6—Automatic machinery soldering leads onto individual solar cells. These machines can replace tedious hand-soldering and produce a better joint.   (Courtesy ARCO Solar, Woodland Hills, California)

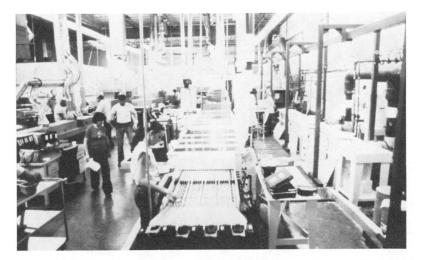

FIGURE 6.7—ARCO Solar's automated assembly line—the first of its kind in the world.   (Courtesy ARCO Solar, Woodland Hills, California)

FIGURE 6.8—Soldering the final connections on a solar cell module. Even in the most automated system, there is still some handwork. (Courtesy ARCO Solar, Woodland Hills, California)

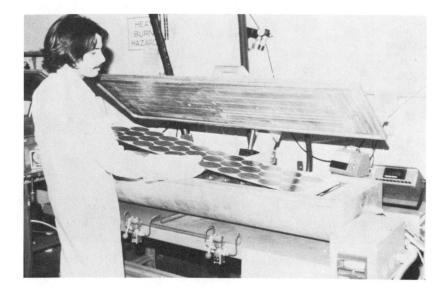

FIGURE 6.9—Sealing the finished module. The best systems are now hermetically sealed in a process similar to that used to make safety glass for car windshields.　　　　(Courtesy ARCO Solar, Woodland Hills, California)

soldering and spot-welding machines are replacing crews that solder cells together by hand. Encapsulant systems with heat-curing plastics are being used instead of silicone resins. And new simplified wire connectors are replacing the expensive junction boxes. The final product is a less expensive but more rugged and reliable device to generate electricity from the sun's energy.

The following chapter describes how new developments promise to carry the state-of-the-art still further, reducing costs by an order of magnitude or more while maintaining the enviable reliability and long life inherent in the photovoltaic cell.

RECOMMENDED READINGS

Chalmers, Bruce. "The Photovoltaic Generation of Electricity," *Scientific American* 235:34–43 (October 1976).

Perez-Albuerne, E. A., and Y.-S. Tyan. "Photovoltaic Materials," *Science* 208:902–907 (May 23, 1980).

Chapter 7

New Developments in Photovoltaic Technology

The standard method of making silicon solar cells is both expensive and energy-consuming. If solar cells are going to compete with the nonrenewable resources now used to generate electricity, new types of cells and new manufacturing processes will have to be developed. The present price of $7 per watt for single-crystal solar cells is, of course, too high to compete with central utilities. The Department of Energy began a research program aimed at reducing this cost, but the recent budget cuts have all but destroyed this program. However, a number of private companies and research groups are still working on a great many alternative approaches to photovoltaic production of electricity. These include new ways of using silicon, other materials, and combinations for solar cells, and clever optical systems to more effectively exploit the photovoltaic process. This chapter will review a number of these systems and will attempt to assess their prospects.

NEW PRODUCTION TECHNIQUES FOR SILICON

Solar-Grade Silicon

The technology for manufacturing silicon solid-state devices has been so well developed, and so much research has been invested in

121

the silicon solar cell, that the most logical way to cut costs is to devise new ways of making this cell. A number of people are doing just that.

Most solar cell manufacturers are small operations relying on hand assembly of arrays, but ARCO Solar (a division of Atlantic Richfield Co.) has set up an automated assembly line to manufacture the cells and assemble them into arrays. The line has greatly increased production capabilities and has lowered array costs to the point where ARCO Solar is selling complete arrays for $9 per watt through their distributors to small users. Direct sales to large projects have included prices as low as $5.50 per watt. Solarex has a great deal of automation incorporated into its Semix production facilities.

The ultrapure semiconductor-grade silicon currently used in the manufacture of single-crystal cells is getting more expensive and more scarce as the electronics industry competes with the cell industry for the limited output of the few producers. The question is, do solar cells require starting material of such high quality? The answer is, not really. A **solar-grade silicon** is now in development that will produce 12 to 15% efficiency solar cells, but that will cost only a fraction of the current $80 per kg paid for the **electronic-grade** used in integrated circuits. Wacker Chemitronic has set up a production facility in the United States to produce nothing but solar cell-grade silicon. Also, Jet Propulsion Laboratory, under contract with the Department of Energy, initiated a research program aimed at developing new processes to produce this grade of silicon in a cost- and energy-efficient manner. This program is being phased out but may possibly have attained the objective of encouraging some manufacturers to proceed with new processes.

Polycrystalline Silicon

One of the most expensive, energy-intensive steps in making a solar cell is the pulling of a large single crystal to serve as the starting material. If a perfect single crystal were not really needed, a great deal of the cost could be saved. Some researchers who have tried polycrystalline silicon believe the loss of efficiency caused by the grain boundaries is just too great, but recently several methods of casting cubes of polycrystalline silicon with large crystal grain sizes have led to polycrystalline solar cells with efficiencies higher than those of the commercial single-crystal cells.

Joe Lindmayer of Solarex started a companion company called Semix, in joint venture with Amoco, to manufacture "semi-crystalline" solar cells from large blocks of silicon cast using a proprietary process. The cooling of the molten silicon is controlled so that the impurities in the starting material are swept to one end of the block, which then can be discarded. This process can proceed using **metallurgical-grade silicon**, currently much cheaper than electronic-grade. The resulting crystal grains in the blocks are so large that the cells produced behave almost like single-crystal cells. If the grain boundaries are perpendicular to the face of the device, the cell acts like a group of small cells in parallel. Semix constructed the Solar Breeder plant to make this new type of cell directly from high-quality sand. The solar cells on the roof are capable of supplying all the electric power needed to run the operation, which is capable of producing enough cells each year to generate 25 peak megawatts of electric power. Unfortunately, since the takeover of Solarex by Amoco, the future of the Breeder is uncertain and the plant is not operating.

FIGURE 7.1—Semicrystalline solar cell.
(Courtesy Semix Inc., Gaithersburg, Maryland)

FIGURE 7.2–Heat exchanger method (HEM) ingot of single-crystal silicon.
(Courtesy Crystal Systems, Salem, Massachusetts)

Crystal Systems has perfected a similar system whereby they can produce a cube of single-crystal silicon 12 inches on a side and 8 inches tall, weighing 90 pounds (see Figure 7.2). Ironically, we are back to single-crystal cells again, but this crystal growing process is very energy-efficient and produces high-quality cells from the metallurgical-grade starting material. The process, which involves a helium gas heat exchanger, casts the ingot in a silica crucible which breaks apart upon cooling, leaving the cube intact. Other organizations, including Wacker Chemitronic, are casting polycrystalline cubes also, but with smaller grain sizes. Kyocera is now selling polycrystalline cell modules made by a proprietary casting process.

Ribbon Growth Systems

The cubes of silicon described above must still be sliced, and, although several groups have developed multiple-wire slicing machines, the process is still very time-consuming and wasteful. A number of people have developed ways to grow long ribbons of single-crystal silicon. These ribbons can be cut into short pieces, and cells can be produced without any further surface treatment.

This method saves several steps and is a much more efficient use of the expensive pure silicon.

Mobil has been working on ribbon growth systems for some time and is in pilot plant production. Their process, which involves pulling a vertical ribbon of silicon out of a carbon die set in a bath of molten silicon, still has problems, however, with inclusions of silicon carbide which form when the hot silicon reacts with the die. (The Japanese, with a similar process, use boron nitride dies to eliminate this problem.)

Mobil's latest ribbon puller produces a nine-sided cylinder rather than a flat ribbon. The continuous molten circumference allows more careful control over the solidification process and production of high-quality sheets with very few silicon carbide

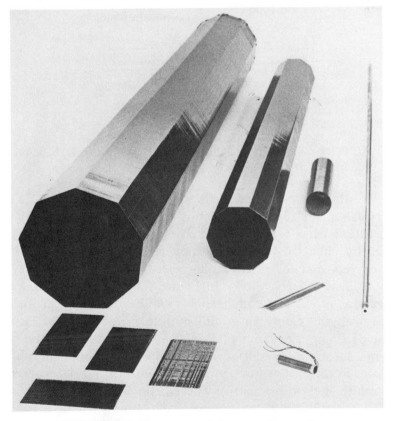

FIGURE 7.3–Nonagon and decagon silicon tubes.
(Courtesy Mobil Tyco Solar Energy Corporation, Waltham, Massachusetts)

inclusions. The cylinder is sliced into 2-inch x 4-inch plates by a high-power laser and solar cells are produced from these plates through a set of proprietary steps. Mobil is now selling competitively priced 40-watt modules made from ribbon growth silicon.

In other ribbon growth methods the ribbon is pulled horizontally, or even downward, or no die is used at all. D. N. Jewett of Energy Materials Corporation has developed a ribbon process that pulls the silicon sheet almost horizontally from a quartz crucible. This system (see Figure 7.4) has the advantage of operating at high pulling speeds while producing a high-quality, defect-free ribbon.

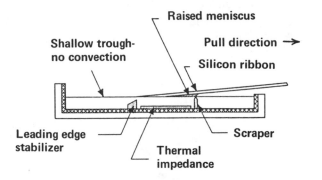

FIGURE 7.4–Silicon ribbon-pulling process developed by D. N. Jewett. (Courtesy Energy Materials Corporation, Harvard, Massachusetts)

A Westinghouse process exploits the dendritic growth of a pair of crystals at the edges of the ribbon to control the shape; Solavolt International has a process whereby a polycrystalline sheet, produced by chemical vapor deposition from silicon compounds, is melted very quickly by a scanning laser to produce a single-crystal ribbon. This process, shown in Figure 7.5, is expected to be in operation producing commercial modules in the near future. Honeywell and Coors, in a joint project financed by the Department of Energy, experimented with ribbons of silicon cast onto ceramic substrates, but all work stopped when government funding ran out. In contrast, a group in France continues similar work with a graphite supporting matrix. The silicon layer can be very thin and if the substrates can be produced cheaply enough, the cell could meet the DOE's cost objective of $0.70 per peak watt, though not by the 1986 target date.

FIGURE 7.5–Laser annealing to form a single-crystal ribbon.
(Courtesy Motorola Semiconductor Group, Phoenix, Arizona)

Other Junctions

Once the thin flat piece of silicon is produced, a junction has to
be formed at the top surface to make a photovoltaic cell. A number
of alternatives to the diffused p–n junction are being investigated,
both to increase cell efficiency and to cut manufacturing costs. It
is possible to make the p and n layers in a solar cell from two different semiconductor materials. This **heterojunction** can have
some real advantages over the **homojunction** made of a single
material. For one thing, the top semiconductor material can act as
a "window," letting more light into the junction and the base
material. Also, a higher junction voltage can be produced if the
materials are selected properly. (For a more complete description
of the heterojunctions and other types of junctions discussed here,
see Chapter 5.)

Tin Oxide and Indium Oxide

Two successful heterojunctions with p-type silicon are **tin oxide** and **indium oxide** doped with tin oxide (called **ITO** by engineers working on the material). Both are n-type semiconductor materials which are transparent to visible light but are good conductors of electricity. Tin oxide can be easily deposited onto glass or other substrates by spraying tin chloride in water onto the heated substrate. Tin oxide-coated glass (called NESA glass) is used as a transparent conductive surface for such applications as aircraft windows. The processes for producing it are well-developed and inexpensive. Indium oxide thin films are a much more recent development, but could possibly end up as easy to produce. However, indium is an extremely rare element and there are some who question the advisability of making solar cells with indium compounds, since such a large quantity of cells will be required to meet a significant portion of our energy needs. In the laboratory, tin oxide and ITO heterojunctions with silicon have been produced which are as efficient as the normal p–n homojunction silicon cells.

Metal–Semiconductor Junctions

Another type of junction that can be used in a solar cell is the **Schottky barrier** or **metal-to-semiconductor junction (MS junction)**. This type of junction is produced when a low work function metal like aluminum is deposited onto a p-type semiconductor or a high work function metal like gold is deposited onto an n-type semiconductor. This barrier contact is in contrast to the low-resistance **ohmic contacts** usually required for metallic contacts to semiconductor devices. If an extremely thin (20 Å) insulating oxide layer is placed between the metal and the semiconductor, the device is called a **metal-insulator semiconductor junction** or **MIS solar cell**. MS and MIS junctions have been made with silicon and nearly all the metals in the periodic table. While some of these devices have shown, in the laboratory, enhanced open-circuit voltages and efficiencies compared to the standard silicon p–n homojunction solar cell, the complexities involved in making the precisely controlled, extremely thin layers required for their successful operation would seem to rule them out as likely candidates for commercial devices in the near future. The research devoted to these junctions, however, has greatly increased our knowledge of how semiconductor devices work and of some of the mechanisms that cut solar cell efficiency.

AMORPHOUS SEMICONDUCTORS

Recently, a whole new class of materials has been investigated for making photovoltaic cells. These are the **amorphous semiconductors**. An amorphous material is one that, while appearing to be solid, has no long-range crystal structure. Glass is a good example of such a material. Amorphous selenium has been known for a century as a photovoltaic cell material and has been used in Xerox copiers since the late forties. The semiconducting properties and the band structure of this material have been examined, and the photovoltaic cells made from selenium have been used commercially for decades even though they have an efficiency of only about 0.1%.

In 1966 Yoshihero Hamakawa at Osaka University in Japan discovered a way of making **amorphous silicon** solar cells. His work was ignored until the mid-seventies when RCA started investigating amorphous silicon as a potential solar cell material. The material had been known for some time but had been dismissed as a solar electric converter because it normally has a high resistance. Actually, if the amorphous silicon thin film is deposited in such a way as to incorporate a small percentage of hydrogen and other elements trapped in the layer, the high dark resistance is immaterial and the solar cell will have a relatively high short-circuit current. It is believed that the hydrogen atoms form Si–H bonds to tie up some of the dangling bonds caused by the somewhat random arrangement of the silicon atoms in the amorphous layer.

The usual way of depositing the amorphous silicon layer is by **sputtering** silane (SiH_4) or some other silicon compound in a vacuum system. The reactive sputtering is achieved by producing a glow discharge in a low-pressure gas mixture using a high-voltage DC or radio-frequency AC power applied between two electrodes. One of the electrodes or a substrate placed between the electrodes will become coated with the amorphous film.

The amorphous silicon cell has now been commercially developed and has spawned an entire branch of the photovoltaics industry. An entire section in the next chapter is devoted to this industry and its prospects.

Amorphous silicon has a band gap a little larger than crystalline silicon and absorbs light more strongly in the visible part of the spectrum, which means that the theoretical efficiency of solar cells made of this material is actually higher than that for single-crystal silicon cells. A great deal of work remains to be done to clarify the semiconducting properties and efficiency loss mechanisms before this efficiency can be realized in practice. However, the simple fabrication steps and the thin films that can be used in making the solar cells should produce a practical, inexpensive cell even if the production line efficiency is no more than the 7% achieved already in the laboratory. Recently, questions have been raised concerning the long-term stability of the amorphous films under sunlight conditions.

Researchers at Osaka University in Japan recently announced a new heterojunction solar cell combining amorphous silicon carbide with amorphous silicon. This device, even though it is very new and little research has been applied to its development, has achieved 7.5% efficiency. It is made by a sputtering system also, but uses a mixture of silane and methane as the reacting gas.

PHOTOELECTROCHEMICAL CELLS

A few years ago, Texas Instruments announced the development of a new type of silicon solar cell. Instead of an all solid-state device with the junction between two different solid materials, this cell has a silicon-liquid junction. The silicon electrode is not in the form of a flat surface at all, but rather tiny beads made in the same way as buckshot. Molten silicon is poured through screens to make tiny droplets that solidify as they fall through the "shot tower." The simplified processing steps and the ability of the device to produce hydrogen as well as electricity had created quite a bit of excitement, and a DOE-funded research program was ordered to develop the idea further. However, when the federal funding ran out, Texas Instruments dropped all work on the device, feeling it wasn't promising enough to invest any more of their own funds.

The **liquid junction** photovoltaic cell is also being investigated in a number of laboratories. Most of these cells use a semiconductor material as the anode and have had problems with long-term stability, as the light-induced chemical reactions can irreversibly change the anode surface. But Bell Labs has developed a new design of liquid junction cell that has the active semiconductor electrode as the cathode and the cell actually becomes more stable when exposed to sunlight. The device can be designed to produce either electricity or hydrogen and steady-state efficiencies of 11.5% have been achieved. However, the cell currently uses scarce indium phosphide as the photocathode material, and any large-scale exploitation of this technique will require the development of a more commonly available cathode material.

CADMIUM SULFIDE/CUPROUS SULFIDE SOLAR CELLS

In the 1960s, Clevite Corporation developed a thin-film solar cell to be used in the space program. The cadmium sulfide solar cell (as it was called) was invented in the early 1950s by Reynolds but it wasn't until 20 years later that scientists figured out that the active semiconductor in the cell isn't cadmium sulfide, but rather the cuprous sulfide layer found when the cadmium sulfide layer is dipped into a copper chloride solution. The cell is another example of a heterojunction solar cell.

The Clevite cell was never very successful because of difficulties in reproducibility of good cells and long-term instabilities in the finished arrays. In the early 1970s, when interest in terrestrial uses of solar energy made a low-cost solar cell attractive, major research on the CdS/Cu_2S cell was again undertaken, mainly at the University of Delaware under Karl Böer. For several years, SES, a subsidiary of Shell Oil, sold a solar array utilizing vacuum-deposited CdS, but it was taken off the market in 1976 due to instability problems. It was found that, because this type of cell can react with the moisture in the air, traces of moist air entered the arrays—even when sealed into glass-fronted enclosures—and destroyed the junctions. SES tried to market the devices several times, but found the sealed modules required were too expensive to produce to market competitively.

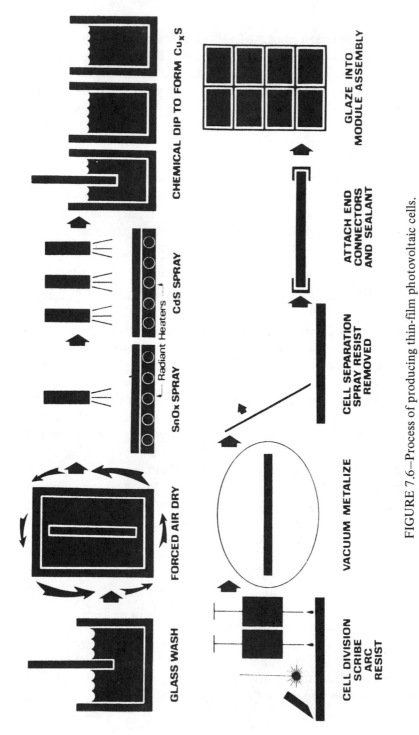

FIGURE 7.6—Process of producing thin-film photovoltaic cells.
(Courtesy Photon Power, Inc., El Paso, Texas)

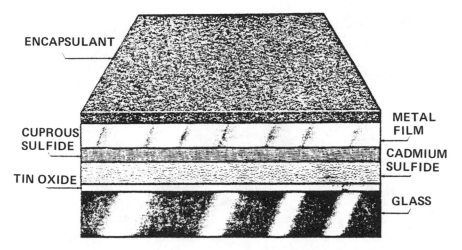

FIGURE 7.7–Diagram of a cadmium sulfide/cuprous sulfide cell.
(Courtesy Photon Power, Inc., El Paso, Texas)

Ease of manufacture from relatively inexpensive starting materials has kept interest alive in this potentially low-cost solar cell. J. F. Jordan, of the Baldwin Piano Company, developed an inexpensive way to spray cadmium sulfide coatings onto glass substrates. His process was taken to the pilot plant stage by Photon Power Corporation of El Paso, Texas, a joint venture between Total, a French oil company, and Libbey-Owens-Ford. Photon Power had expected to sell 4 to 5% efficiency solar arrays at $2 per watt of capacity when they went into full-scale production, but startup problems caused them to drop the project. Recently, Photon Energy Corporation has been put together from the remnants of Photon Power and has been making large-area (4 ft^2) electroplated cadmium telluride cells with some assistance from the Solar Energy Research Institute (SERI).

Some questions have been raised about the safety of CdS/Cu$_2$S solar cells because cadmium is a poisonous heavy metal. Of course, it is extremely important that introduction of a new technology not be a cause of new environmental problems when applied on a large scale. It is ironic, however, to compare the accessibility of cadmium in solar cells to that used to plate refrigerator racks and the bolts on children's toys. A more important problem might be the relative scarcity of cadmium compared to the quantity that would be needed for full-scale exploitation of thin solar cell systems. It was recently discovered that the addition of zinc to cadmium sulfide markedly improves efficiency. Ten percent efficiency has been realized in the laboratory.

FIGURE 7.8—Town of Paradise Valley, Arizona, police repeater on Mummy Mountain.   (Courtesy Solavolt International, Phoenix, Arizona)

There are a number of materials similar to cadmium sulfide and cuprous sulfide that can also be used to produce photovoltaic junctions. These II-VI compounds (named after the second and sixth columns of the periodic table) are being investigated and several promising materials such as cadmium selenide and cadmium telluride have been used in solar cells. These materials can be deposited as thin films by essentially the same processes that are used for cadmium sulfide films and also by electroplating. Ametek, Incorporated of Harleysville, Pennsylvania, has electroplated N-I-P cadmium telluride solar cells of 11% efficiency, a new record for this material. Boeing Corporation has replaced the cuprous sulfide layer with one of cuprous indium telluride and has achieved a thin-film cell with over 10% efficiency. While a great number of combinations are possible, the development of these semiconductor materials is restricted by a limited research effort. The government's decision to slash photovoltaic research funds makes this situation even worse.

III-V SEMICONDUCTOR COMBINATIONS

The III-V group, a whole class of compound semiconductor materials, derives its name from the third and fifth columns of the periodic table. The compound semiconductors are made by choosing one (or more) element from each column. Gallium arsenide is the best known of these substances, but the list also includes gallium phosphide, indium phosphide, and more complex mixtures such as aluminum gallium arsenide. The III-V compounds are important commercially; they are used in making the light-emitting diodes in pocket calculators, for example.

The III-V semiconductors have been studied extensively, their properties are well understood, and they can actually be tailored to specific applications. The band gap of gallium arsenide is almost ideal for the construction of high-efficiency solar cells, and the high-quality crystals made of these materials allow most of the light-generated current to be collected by the contacts. The material will also withstand surprisingly high temperatures without damage. Consequently, the $\sim$ 25% efficiency attainable with these solar cells represents the greatest efficiency of any solar cell system produced so far. Varian Associates is working on extremely high-quality III-V compound solar cells to be used at the focus of high-power concentrating systems. The only problems with these cells are cost and availability.

All the III-V compound solar cells developed so far use gallium and/or indium. Both elements are extremely rare, much more so than gold, and the only reason the prices of these metals are not astronomical is that, until recently, there was essentially no commercial use for them. They are produced as by-products in the smelting of other metals, notably aluminum and zinc. Also, the processes presently used to produce the single-crystal materials, such as epitaxial growth, are expensive and energy-intensive. The only reasonable way to generate large amounts of solar power with such cells is to use very small amounts of these scarce elements per watt of capacity. Two possibilities are thin-film cells and concentrator systems, such as the Varian development mentioned above. It is impossible to tell if these cells will make a large contribution to the solar cell industry, but they will certainly be used in those situations where considerations of high efficiency or good thermal stability outweigh those of cost.

One use of photovoltaics in which cost is no object is military space hardware. The CIA is planning to use gallium arsenide cells in their satellites, and the Department of Defense is sponsoring research to investigate promising III–V compound solar cells, possibly to power "Stars Wars" weaponry. These grants, small by military standards, are becoming a large portion of the support the U.S. government is giving solar cell research.

ELECTROPLATED CELLS

Many of the new semiconductor materials are extremely strong absorbers of light. They are called **direct band gap semiconductors** and the light that is absorbed to create the charge carrier is absorbed in the first micron or so of the material. This means that the entire solar cell can be extremely thin and can be made of polycrystalline material if the average grain size is larger than the micron or less that the charge carriers have to travel. If the grain boundaries can be **passivated** so that they won't act as carrier recombination centers, the crystal grains can be even smaller. (A **recombination center** is a point at which a hole–electron pair created by the absorbed light can recombine before the electron passes through the external circuit. It acts as an internal short circuit.) With thin layers of small grain size, electroplating is an excellent way to build a thin-film solar cell. Electroplated cells have been made with cuprous oxide, zinc phosphide and cadmium telluride.

Cuprous oxide is called an emerging material by the planners in the Department of Energy even though it was the first known photovoltaic cell material and was studied in the early part of this century. It is a p-type semiconductor and simple Schottky barrier solar cells can be made by simply heating a sheet of copper in air for 15 minutes to 1000°C, followed by annealing at 500°C for another 15 minutes. This solar cell is extremely inefficient because the active junction is between the thin cuprous oxide layer formed and the underlying copper; the light must pass through the entire cuprous oxide layer to get to the junction. Several years ago Dr. Dan Trivich and I developed a method of making front wall

Schottky barrier solar cells in which the active junction is between cuprous oxide and a thin, semitransparent metal layer covering its surface. These cells display an efficiency of only 2%, but our theoretical studies indicate that an efficiency of over 13% may be possible, if the right combinations of materials and processing steps could be found. It is possible to electrodeposit cuprous oxide thin films in a process that would cost pennies per square foot; if cells of only 5% efficiency could be thus manufactured, costs of $0.25 or less per watt of capacity would be achievable. The supporting substrate would be the most expensive item in the finished product.

Similar costs per watt would be possible for electroplated cadmium telluride solar cells also. One system, now being developed by Ametek, has announced 6% efficiency in the laboratory. However they are not releasing details of their proprietary process and do not expect commercialization of the process for several years.

Many combinations are possible and a number are already under investigation on a small scale. The main barrier to developing inexpensive, efficient solar cells is a lack of knowledge of the semiconducting properties of these promising materials.

ORGANIC SEMICONDUCTORS

Among the little-explored potential solar cell materials are organic semiconductors. These materials have been used in experimental photovoltaic cells in two ways—as photosensitizers and directly as the semiconductor element in a cell structure. It has been known for a long time that certain classes of dyes can, if deposited in a thin layer onto a base material, make that material photosensitive to the light the dye absorbs. This technique has been used for nearly a century now by the photographic industry to make orthochromatic and, later, panchromatic film. The silver halides used in photographic film (silver chloride, bromide or iodide) are only sensitive to blue or ultraviolet light, but the dyes used to coat the tiny crystals of silver halides can absorb the rest of the visible spectrum, and somehow transfer the energy of the absorbed light to the underlying crystal. It is now generally believed that this transfer involves the actual transfer of a charge carrier (either a hole or an electron) to the substrate material. The same technique has been commercially used in copying machines.

Amal Ghosh at Exxon, and others, have succeeded in making photoelectrochemical cells that can produce both hydrogen and electricity by dying titanium dioxide-coated electrodes with the organic dye phthalocyanine, thus rendering them sensitive to the visible sunlight that titanium dioxide (the main constituent of good quality white paint) normally reflects. The efficiency of such cells is still low, but they appear to be quite stable and would be very expensive to produce.

Organic semiconductors have also been used directly in Schottky barrier photovoltaic cells. About 20 years ago, while experimenting at Xerox Corporation with layers of phthalocyanine and other organic dyes in an attempt to determine the mechanism of energy transfer between the dyes and various substrates, I accidentally discovered that some of the devices exhibited a photovoltaic effect. We studied this effect for a while since it looked promising as a photodetector for Xerox's high-speed facsimile system then under development, but nothing was done at that time to exploit the system for solar energy conversion. Recently, Xerox of Canada has taken up the idea as a potential solar cell, but the efficiency achieved so far is still low because of the poor conductivity of the material. But this class of organic pigments, all similar to chlorophyll in chemical structure, could produce an extremely cheap solar cell that could be literally painted onto the substrate.

There are a number of other dark horse solar cell possibilities that have been given only cursory research efforts. Any one of these, if proven to be commercially useful, could suddenly change photovoltaics from something used only in remote areas and for special situations to the system that replaces all conventional utilities. Chlorophyll membrane systems, photosensitized zinc oxide, and little understood compound semiconductor systems would all fall in this category.

CLEVER OPTICAL SYSTEMS

In addition to exploring new solar cell materials, researchers are also looking hard at ways of making existing solar cells utilize sunlight more efficiently. Most of these systems employ clever optics to direct or focus the light before it is absorbed by the cell.

The simplest systems are the hybrid optical concentrator systems discussed in Chapter 1. New inexpensive fresnel lens optical systems, nonfocusing concentrators that allow large tracking errors, and automated assembly systems promise to lower the final installed cost of concentrator systems to make them competitive with the cheapest flat plate photovoltaic collectors. Most of these new designs incorporate methods to extract the heat generated in the solar cell and put it to use, making the system even more economically attractive. Sunwatt has just introduced an inexpensive hybrid concentrator based on an aluminum fin with a copper tube, similar to the design of a solar water heater.

CASCADE CELL

The cascade cell, first proposed in 1953 by Trivich, is simply a stack of different solar cells. Each one absorbs part of the solar spectrum and passes the rest of the light through to the other cells in the stack. The cell made with the semiconductor which has the largest band gap has the shortest cutoff wavelength and transmits most of the light redder than this wavelength. This light can be utilized by a semiconductor with a narrower band gap placed behind the first cell. It is possible to stack even a third cell or more and efficiently use the entire spectrum of sunlight including the infrared. A single, narrow band gap cell would also absorb the entire solar spectrum, but would have a low output voltage, essentially wasting a good deal of the energy of the short-wavelength light. The cascade cell will have an output voltage that is the sum of the voltage of the individual cells. The output current of the device is less than that of the simple, narrow band gap cell, but the overall efficiency ends up much higher. Experimental two-cell devices have been constructed using gallium arsenide and aluminum gallium arsenide and various combinations with silicon, gallium arsenide and germanium. In some of these systems all the various cell layers are constructed directly on top of each other; they utilize a **tunnel junction** or some other sort of shorting junction to make an electrical connection between the cells in the device. Actual efficiencies of over 23% have been measured and theoretical efficiencies of 40% or more have been predicted for these multijunction devices.

The Department of Defense is interested in these **tandem cells** because of the great amount of power delivered by a small package. A photovoltaics-powered unmanned airplane, with two-layer tandem cells covering the wings, is now a possibility. The cells would produce enough extra power, stored in the batteries, to keep the plane flying all night. The drone could circle for months at high altitude, taking surveillance pictures.

Most experimenters envision using the cascade cell at the focus of an optical concentrator system since the cost of the complex cell is expected to be high, and the efficiency of the system actually increases with greater light intensity.

Another version of this is shown in Figure 7.9. Here the incoming light is split into various colors with selective filters and directed to several different cells, each of which is designed to use a particular band of wavelengths more efficiently. Varian Associates has also been working on a version which uses high concen-

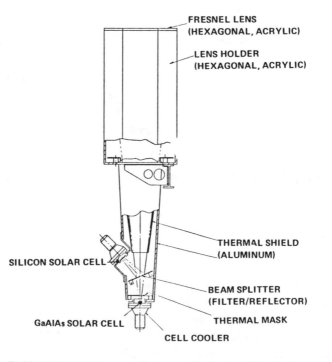

FIGURE 7.9–Diagram of a two-cell photovoltaic device using concentrated sunlight.

tration ratios and very efficient dielectric-coated filters as the beam splitter. Another, potentially cheaper version being developed by W. H. Bloss at the University of Stuttgart, uses holograms developed in gelatin as the dispersive concentrating (DISCO) system. The DISCO system focuses the light and splits it into several bands of different wavelengths directed at the two or three different solar cells used.

LUMINESCENT CONCENTRATORS

One of the most exciting new concentrator systems is the luminescent solar concentrator, which absorbs solar radiation in a flat plate by means of colored dyes. These dyes fluoresce in all directions at a longer wavelength than the incoming radiation. Most of this reradiated light is confined to the collector by total internal reflectance and transmitted to the edge of the sheet where small photovoltaic cells convert it into electricity. The basic concept has been known for years and, in fact, has been used in drafting triangles for illumination, but the application to solar energy collection was first proposed in 1976 by Lambe and Weber. Since then, a number of groups have started work on the idea, using organic dyes in plastics or inorganic fluorescent compounds in glass. A rather complete theoretical treatment of the efficiencies and concentration ratios to be expected is given by Goetzberger and Greubel.

Wood and Long experimented with liquid-filled flat plastic cases as luminescent concentrators in which a mixture of dissolved dyes absorbs most of the visible solar spectrum. A cascading of energy, which they envision as a series of fluorescences and reabsorption by the different species of dye molecules, produces a final fluorescent light which is transmitted at a long wavelength to the photocells at the edges of the case. In their experiments they found a rapid heat buildup in the liquid as all the incoming light energy that is not finally converted to electricity or escapes from the case is trapped as heat in the system. They also found that the dyes bleached rather quickly under their experimental conditions.

At Skyheat Associates, I have been working on a version of a luminescent concentrator that utilizes both the heat and the

FIGURE 7.10–A luminescent concentrator. This flat plastic tank contains a
colored fluorescent liquid. The liquid concentrates light onto small solar
cells glued to the edges of the tank (Skyheat Workshop photo).

electricity developed from the sunlight. Dyes dissolved in a liquid
absorb the solar radiation and transfer the light to solar cells
fastened to the end of the flat tank. The liquid is heated and can
be pumped through a heat exchanger to extract the heat generated. Cooler operating temperatures and a careful choice of dyes
seem to greatly reduce the dye stability problem, but the system is
still in early development and working models are too inefficient
to be of practical interest. At present, the only research being done
on luminescent concentrators–at Owens-Illinois–is winding up.
Funds have been cut completely in this promising area.

THE SOLAR THERMOPHOTOVOLTAIC SYSTEM

This complicated sounding name describes a system that com-
bines some of the aspects of solar thermal concentrator devices
with a method of converting directly into electricity the heat
generated by focused sunlight. If a large parabolic mirror focuses
sunlight through a small opening into a well-insulated black
cavity, the interior of the cavity can be made extremely hot
(1500°C or more). This temperature causes the inside walls to

radiate in the infrared, with the radiation bouncing back and forth in the cavity. If a germanium photovoltaic cell is placed in one wall of the cavity, the infrared absorbed can be converted into electricity because of the narrow band gap of germanium.

Loferski and others at Brown University are developing a special, high-efficiency germanium solar cell especially for this application and calculate that overall conversion efficiencies as high as 20% are possible with such a system. The germanium cell would actually stay cool, even in such a hot environment, since the incoming infrared energy is converted into electricity. However, if the cell were disconnected from the electric load, it would quickly heat up and destroy itself. The solar thermophotovoltaic system has the disadvantage (or advantage, depending on your viewpoint) of only being practical in a large centralized system like a **power tower** where economy-of-scale justifies the expense of the complex tracking system for a whole field of steerable mirrors.

RECOMMENDED READINGS

Ariotedjo, A. P., and H. K. Charles. "A Review of Amorphous and Polycrystalline Thin Film Silicon Solar Cell Performance Parameters," *Solar Energy* 24:329–339 (November 1979).

Goetzberger, A., and W. Greubel. "Solar Energy Conversion with Fluorescent Collectors," *Applied Physics* 14:123–139 (1977).

Hovel, Harold. "Photovoltaic Materials and Devices for Terrestrial Solar Energy Applications," *Solar Energy Materials* 2:277–312 (1980).

Proceedings of the IEEE Photovoltaic Specialists Conferences, May 11, 1975, November 12, 1976, June 13, 1978, January 14, 1980, May 15, 1981, September 15, 1983, May 1, 1984 (New York: IEEE).

Chapter 8
The Photovoltaics Industry

A young, rapidly growing solar cell industry now exists. Product lines have been standardized, and in 1983 companies shipped 20 megawatts worth of photovoltaic modules worldwide. A secondary market in **balance of systems** equipment has emerged to furnish the nonsolar cell parts needed for a complete installation. A number of firms have begun to specialize in designing and installing finished systems in remote sites all over the world.

THE BOOM IN AMORPHOUS SILICON

One of the most surprising industrial developments has been the rapid growth of the amorphous solar cell. From a laboratory curiosity a few years ago, the cell has become a standard part of pocket calculators and watches. (Industry sales amounted to 60 million calculators in 1984, a substantial part of the total Japanese cell output.) An extremely sophisticated, automated manufacturing process has reduced the price of a set of four or five small cells to less than the cost of the batteries they replace. Even though the cost per watt of small amorphous cells is around $30, the milliwatt power cells in use are only a few cents each.

Two obstacles to the immediate large-scale production of amorphous power modules are light-induced degradation and poor yield in large-area cell production.

The first of these problems, the irreversible deterioration of the cell after exposure to direct sunlight for as little as one day, is poorly understood, even though a great deal of research effort is now directed toward the phenomenon. The extent of decline has been reduced by changing the formula of the complex mixture of gases used in the sputtering deposition of the amorphous layer, but the effect is still present and, for now, simply accepted. Some manufacturers refer to this as the "burning in period" and rate performance after degradation. The relatively free movement of small atomic species, such as hydrogen and fluorine, in the loosely linked silicon atom matrix could be responsible for the light-induced degradation; a good deal of research effort is aimed at replacing the fluorine and "locking up" the matrix with larger atoms such as other halogens and boron. Boron, of course, also acts as a sort of p-type dopant, giving the amorphous film the proper electronic characteristics for a solar cell. The research currently underway worldwide will certainly pay off in the near future and we will soon see commercially available amorphous solar cells of 6 to 8% efficiency.

The second problem, poor yield in large-scale manufacture, could simply be one of quality control similar to the problems that plague all semiconductor manufacturers when they enter a new product area. It is conceivable, though, that the point defects that could short out an entire cell will be almost impossible to eliminate. In this case, the best strategy might be to institute an "integrated circuit" approach—large amorphous photovoltaic modules composed of thousands of tiny cells all made at one time on a single substrate, connected only after testing has revealed the defects.

It is becoming difficult to keep track of the corporations that have announced new or expanded efforts in the amorphous solar cell area. The Japanese are the clear leaders at this time and, either directly or through joint ventures with American companies such as Energy Conversion Devices, are responsible for most current commercial sales. Japan has built demonstration homes with amorphous power panel roofs, and several companies have shown prototype power panels in the 30- to 40-watt size. None of these companies accept orders for immediate delivery, and schedules are vague and have a tendency to slip. Actual full-scale production will have to wait until the above-mentioned problems are solved.

In the United States, Energy Conversion Devices (ECD) and Chronar depend on amorphous cells for their entire product line. ECD, in conjunction with Sharp, maintains a full-scale automated production line in Japan to fabricate a wide ribbon of cells on a continuous basis; two more lines will be operable soon in the U.S. and there are plans to export the technology on a turn-key basis to developing countries. While ECD has demonstrated large-sized, flexible amorphous modules, their production modules are small units for calculators and consumer devices that are not exposed to long hours in the direct sun.

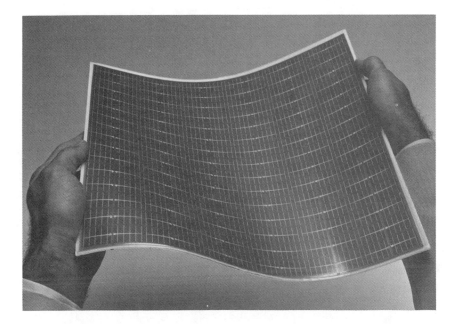

FIGURE 8.1—One-foot-square Ovonic amorphous silicon flexible cell.
(Courtesy Energy Conversion Devices, Inc., Troy, Michigan)

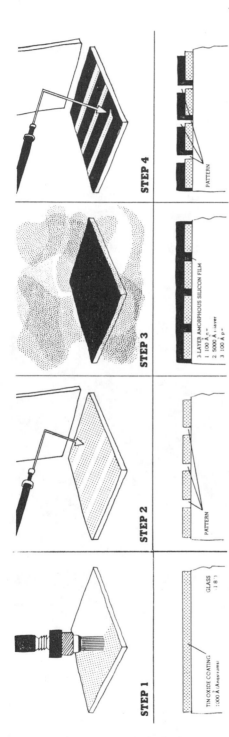

STEP 1

TIN OXIDE COATING
:000 Å (Angstroms)

GLASS
1 8 1

The glass substrate is coated with a transparent conductive coating. Tin oxide is deposited by a chemical spray deposition process from organometallic tin compounds.

STEP 2

PATTERN

The transparent conductor is patterned into regions of conductive and nonconductive areas by removing the tin oxide coating with a scanned high power laser.

STEP 3

3 LAYER AMORPHOUS SILICON FILM
1 100 Å n⁻
2 5000 Å i-layer
3 100 Å p⁻

A deposition chamber lays a film of amorphous silicon on top of the patterned tin oxide. Three different layers of amorphous silicon are deposited: one layer with p-type conductivity, a middle intrinsic layer, and a top layer with n-type conductivity. This configuration converts sunlight into electricity.

STEP 4

PATTERN

Strips of the amorphous silicon films are removed, using laser techniques similar to Step 2.

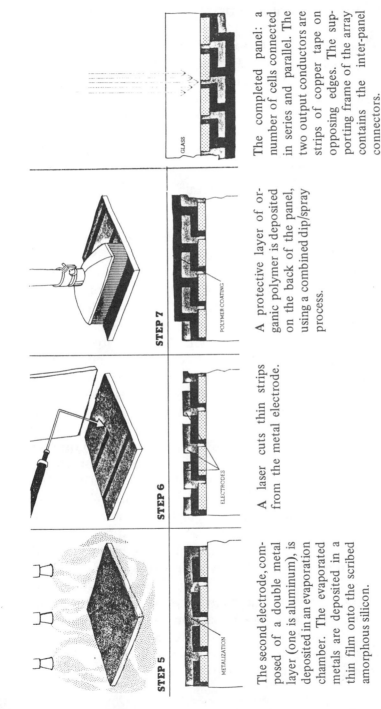

STEP 5

STEP 6

STEP 7

METALIZATION

ELECTRODES

POLYMER COATING

GLASS

The second electrode, composed of a double metal layer (one is aluminum), is deposited in an evaporation chamber. The evaporated metals are deposited in a thin film onto the scribed amorphous silicon.

A laser cuts thin strips from the metal electrode.

A protective layer of organic polymer is deposited on the back of the panel, using a combined dip/spray process.

The completed panel: a number of cells connected in series and parallel. The two output conductors are strips of copper tape on opposing edges. The supporting frame of the array contains the inter-panel connectors.

FIGURE 8.2.—Amorphous silicon photovoltaic panel manufacture: the Chronar process step-by-step. The process is capable of handling glass up to one foot by three feet. By varying the laser scribing pattern of the various layers, different combinations of voltage and current can be obtained. (Courtesy Chronar Corporation, Princeton, New Jersey)

Chronar, on the other hand, has concentrated on large-wattage power modules from the start and consequently is slow to sell solar products on the open market. So far, Chronar's profits have been produced by selling technology: arranging joint ventures in Egypt and elsewhere to set up amorphous assembly plants. Their own plant, built in conjunction with AFG Industries, Inc., uses automated technology to build amorphous arrays onto tin-oxide-coated glass (see Figure 8.2). These small cells, produced for use in calculators, run at about 3% efficiency. Chronar has also introduced an innovative solar patio light, utilizing a 2-watt amorphous silicon panel which charges a battery to operate a small (0.8 watt) light. These inexpensive outdoor lights have been selling very well, giving Chronar a profit for the first time, in the second quarter of 1988.

Most major solar cell manufacturers have announced large amorphous silicon research and development efforts. Some companies, like Exxon and Photowatt SA, have closed their crystalline silicon operations, pinning all their hopes on a future amorphous solar cell line. Others, like ARCO Solar and Solarex, plan to phase in the amorphous products as they are ready, but are not decreasing conventional solar module production. Solarex, for example, bought RCA's amorphous silicon research and development section. In fact, it was the pioneering work at RCA, the first American company to work on amorphous silicon cells, that lead to this recently developed industry. Even though the basic cell was discovered in a Japanese university laboratory, Japanese companies paid no attention to the work until RCA announced its amorphous program at the 1976 IEEE Photovoltaic Specialists Conference.

The boom in amorphous silicon solar cells will continue. It is only a matter of time before the technical problems will be solved and large power modules become generally available. The price per watt for these devices, however, will not drop precipitously. The cost of manufacture and the high reject rate will make it difficult, at first, to profit, especially in competition with the present price per watt of crystalline silicon cells. The price will start to decline only after large-volume deliveries have given production people the time and experience to take advantage of the inherent economics of amorphous devices. As this happens, new markets will open up, and amorphous photovoltaics will begin to produce a significant amount of our electric power.

THE OIL COMPANY–SOLAR CELL CONNECTION

The majority of companies making solar cells are now owned by oil companies. There is considerable debate whether this is good for the industry and good for the ultimate consumer. Before we condemn the trend outright as another example of monopolizing a good thing, we should examine it in detail.

The multinational oil companies see themselves as energy companies and, after gaining control of most coal and uranium mines, a natural step would be to move into solar energy. This move has not been without difficulties; easily attainable, centralized resources do not exist in the solar field. The inherent decentralization of solar and most renewable energies does not fit into oil companies' existent distribution and marketing structure. Since solar energy must compete with the well-established and heavily subsidized conventional energy sources, the profit margins are very slim—a new experience for large oil companies. Exxon, for example, sold a solar water heater company it had acquired because neither the company nor the market met expectations.

Why, in the face of these difficulties, would the oil companies want to control this industry? As our strong dependence on petroleum is forced to an end, the oil companies must cover all the options if they are to continue as the world's energy suppliers. Having acquired large interests in the fossil fuels and in nuclear power, it is easy, given the size of the solar cell industry and the capital resources available to large oil companies, to buy fledgling solar cell firms. The oil companies know that the rising cost of conventional energy, the present delivery prices of photovoltaic cells, and the promise of a breakthrough in solar cell technology will lead to an enormous growth potential for photovoltaics, and they want to participate in that growth.

The real question is, is this good or bad? Is it helpful or harmful for a few large companies to control a world energy source? And will it be good for the solar cell companies to be run as small parts of giant corporations?

The theory that oil companies buy solar cell companies to put them out of business and make photovoltaic cells unavailable is unrealistic. This is impossible to do. Solar cells have already established themselves as so indispensible in certain applications, *e.g.,* space satellites, that their manufacture must continue. In addition, the technology for making photovoltaic cells is so well-established that suppliers to the terrestrial market can enter the field faster than oil companies can buy them out. There is no evidence that the solar cell companies already owned by oil companies are not attempting to produce better and cheaper solar modules and to expand their markets. Occasionally, an entire company's product line will disappear, as did Exxon's Solar Power products. Rather than attempt to make solar cells unavailable, this is a demonstration of the power of unlimited capital resources. A giant corporation can simply stop selling a product for years, if necessary, and absorb the operational expenses of an entire division while developing a new product line and production facilities. Besides, these losses offset the profits that are subject to the windfall profits tax.

The availability of large amounts of research and development money for new photovoltaic systems and the capital resources to quickly scale up production with automated assembly lines could be a boon to the solar cell industry. The government decision to slash federal funding for solar cell R&D makes the cash available from oil companies even more desirable and, some claim, necessary. So it is possible that, given the present situation, oil company ownership of the solar cell industry could do a great deal of good. They have the resources to expand the industry much more rapidly than the small companies that must bootstrap themselves on the slim profits presently available. Venture capital for a new technology like photovoltaics is difficult to obtain from conventional sources at present; the oil companies do have the capital and they do appreciate the importance of photovoltaics.

Since the first edition of this book was published, the trend of acquiring companies is continuing. Solarex has been completely bought out by Amoco (Standard Oil of Indiana). Amoco has had a minority position in Solarex for years but, in 1983, circumstances led Solarex to cash flow problems, and the oil company took the opportunity to acquire the remainder of Solarex stock at bargain price. Two events led to the takeover. First, when RCA decided to give up its amorphous silicon research and development, Solarex saw the chance to acquire the leader in the field, which they correctly perceived as the biggest potential threat to the Semix polycrystalline silicon process. Second, Solarex had just borrowed heavily to construct the Solar Breeder in Maryland. This innovative solar cell factory, powered by a roof covered with solar cells, gave Solarex the capability to make photovoltaic cells—starting from sand—without any outside power source. Unfortunately, at the time the Solar Breeder was finished and dedicated, sales of solar modules were down and the cash flow was not sufficient to pay off the loans. Amoco supplied the cash and promptly shut down the plant.

The Solar Breeder apparently works well but has never produced any modules. Recently I asked Solarex to quote the purchase price of 50 Semix solar modules for a self-sufficient home project and was told there was a several month delay in delivery because the current production facilities couldn't keep up with the increased demand. Perhaps it is time for the Solar Breeder to be put into operation.

A small photovoltaic subsidiary company with a strong financial commitment from the parent company can engage in new and innovative research. A good example is the decade-long commitment of Mobil to the ribbon-growth process for making single-crystal silicon solar cells. The project started as a joint venture with Tyco but is now solely owned by Mobil. Mobil's Ra photovoltaic modules were the primary electrical source, covering a good part of the roof, of a home featured on PBS' "This Old House." Mobil recently took the panels off the market, but R&D work continues. The panels were quite expensive; a breakthrough in technology does not always lead to an immediate breakthrough in price.

A number of other oil companies invested in photovoltaic technology. SOHIO joined with Energy Conversion Devices to build several amorphous solar cell production lines ("Sovonics"). Chevron involved itself in high efficiency tandem solar cells, and Diamond Shamrock started a small advanced photovoltaics R&D effort.

Shell has been in the photovoltaics field for quite a long time; they started a company called SES to develop cadmium sulfide/cuprous sulfide solar cells. Then they became the controlling partner in Solavolt International, formed as a joint venture with Motorola. After several failed attempts to achieve the required reliability from CdS/Cu_2S cells,

SES was disbanded and all effort went to Solavolt's silicon ribbon process. When that process was ready for full-scale production, Shell laid off the research and development staff. This is a strange decision as industrial experience shows the need for research people is greatest when a product moves out of the laboratory into the plant. Things never seem to work right at first and someone has to figure out why.

A disturbing recent trend is the closing of solar cell divisions by the parent oil companies. In addition to Shell's decision, which is really a re-alignment of product type, Total and Libbey-Owens-Ford completely abandoned their Photon Power CdS/Cu_2S project. The closings of Solar Power by Exxon and Photowatt by Elf-Aquitaine created the biggest news. Both were major suppliers of photovoltaic power modules and represented a major share of the industry's capacity. The parent companies blamed the slow growth of the industry for the closings. This, coupled with the low selling price of photovoltaic modules as compared to their cost of manu-facture, has kept the major photovoltaic companies from making any profits. (Both parent companies have, however, kept open the option of producing amorphous silicon cells.)

ARCO Solar has been the leader in price cutting. In 1979 and 1980, ARCO sold solar modules at a price of $7 per watt to several big buyers, including the federal government. This was well below the cost of production at that time. Large-scale production plus new automated facilities have lowered the labor and materials cost for 35-watt PV modules to around $5 per watt, but recent bulk purchases at $5.40 per watt leave no room for overhead or profit. Such a pricing policy forces out the competition, especially those small, independent companies with insufficient financial backing. The few independents left survive by concentrating in areas not currently of interest to the oil companies. Applied Solar Energy Corporation is profitably producing space rated cells and cells to be used in high-ratio concentrator systems. Solec and Solenergy both sell individual cells to small assemblers who work on specialty products, until recently a market ignored by the oil companies.

One large market for solar cells is the oil companies themselves. They know what it costs to furnish electric power to remote sites and now use photovoltaic arrays on oil drilling rigs (see Figure 8.3) and for corrosion protection devices on pipelines. A parent oil company could probably use the entire output of its solar cell division, giving incentive for the division to expand and produce greater quantities of inexpensive solar modules. The division, it is hoped, would continue to sell to the general public and pass along the cost savings.

FIGURE 8.3—Photovoltaic modules are being used by oil companies on offshore drilling rigs to operate the marker lights and communication equipment. (Courtesy Solenergy Corporation, Woburn, Massachusetts)

It is impossible to predict the long-term results of the oil company—solar cell connection. Given the resources, solar cell companies could grow and become a very important part of the energy industry, furnishing virtually all our electric power needs at a reasonable price. The nagging concern is the advisability of allowing this important energy source to be owned by the same organizations that control our other energy sources.

PHOTOVOLTAIC POWER FARMS

Ever since the nineteenth century when the amorphous selenium cell was developed, people have dreamed of photovoltaics' potential as a major electric power source. Now with the existence of **solar power farms**—large racks of solar cells that generate electric power in remote areas and then transmit it to the power companies—the dream is being realized.

The first large-scale photovoltaic power systems were United States government demonstration projects: Indian villages in the southwest, an irrigation system in Nebraska, even a radio station in Ohio were powered from arrays that represented a major portion of all the solar cells produced in the mid-seventies. These demonstration projects gave the industry a needed boost and produced a database of operating experience and module failure modes that has proved invaluable in the search for more reliable designs.

Large solar cell power arrays were next applied in areas where conventional power was unavailable and the cost of extending the utility grid prohibitive. Villages in Saudi Arabia and Africa were fitted with these experimental systems, each one a custom design. The Saudi system is particularly interesting in that it incorporates long racks of Fresnel lens concentrators to focus the sun onto specially designed photovoltaic cells. The experience in designing the controlling and tracking mechanisms has lead to new levels of sophistication in electronic circuits and drive mechanisms. The array has a peak output of 350 kW.

All these projects were subsidized at first by a government. It was believed that it would be years before photovoltaic cells became cheap enough to justify their large-scale use in competition with more conventional generating stations as part of a utility grid. But the passage of the Public Utilities Regulatory Powers Act (PURPA) by the U.S. Congress changed the economics. It is now possible for a small producer (less than 80 megawatts) to install a generating system and sell the power to the utility at a favorable price and without the enormous amount of red tape usually required of a new electric power producer. This new act plus the renewable energy and investment tax credits have made renewable energy attractive as an investment. In California, wind power farms are showing up in every pass or windy spot capable of

FIGURE 8.4—350-kW photovoltaic power farm is furnishing power for remote villages in Saudia Arabia. Martin-Marietta designed and built the tracking concentrator arrays.   (David Buzby photo)

handling the installation of wind generators, and small hydro-electric plants are being built along almost every suitable stream. Investors have included photovoltaics in their plans, and solar power farms are now a reality.

In December 1982, the first of these photovoltaic power farms went into operation near Hesperia, California. The one-megawatt power station, built entirely by ARCO Solar, incorporates an unusual set of tracking units that allow a set of flat plate modules to follow the sun for about eight hours every day. No concentrators are used in this system.

ARCO Solar also won the bid to furnish the first sections of a very large photovoltaic power plant being built by the Sacramento Municipal Utility District. The 12-year project is expected to furnish at least 100 MW when finished and should produce power for less per kilowatt hour than the nearby Rancho Seco nuclear power plant. Interestingly, while the Sacramento utility had no problem getting single-crystal and polycrystalline silicon cell modules at low prices, no amorphous silicon manufacturers bid or promised definite delivery dates. The Sacramento utility project is being financed partly through the Department of Energy, money they intend to repay when the average price per watt of modules in future purchases becomes less than $3.20.

Numerous solar power farms were rushed to completion before the solar tax credit expired. It was hoped that by the time this happened, the price of solar cell modules would have dropped to the point where investment in large central systems made good sense without the artificial support offered by the tax credits. Although the price did not drop to the extent hoped, solar cells have become cost-effective in many applications.

The Electric Power Research Institute (EPRI) has issued a series of reports over the years evaluating photovoltaics as a potential electrical generator. Though they only consider systems from a utility's point of view—dismissing distributed solar cell arrays on individual homes, even those that would be grid-connected—their study of the economics of the centralized solar power farm is informative and carefully done. They conclude that solar cells have to be both cheap and efficient before a utility could justify considering them as an alternative to coal or oil. Nuclear power plants were not mentioned in this report; their economics compare quite unfavorably, of course, to the alternatives. Because of the very high balance of system costs assumed for the power farms, the conclusion is that only advanced cell types or concentrator systems have any chance of economic acceptance. In spite of the pessimistic view of the utility industry, plans are going ahead for more and larger photovoltaic power farms. Most of these will use flat plate modules, but some will incorporate concentrators and cogeneration.

In the future, the photovoltaic roof will become more common on homes that have become part of a distributed solar power farm, saving a good part of the balance of system costs related to the acquisition of land and building of support racks needed to hold the thousands of solar cell modules required. These distributed power farms can be organized and financed in three ways:

1. The individual homeowner may decide to invest in an array, expecting the savings in utility bills and the small return each month from the utility to eventually pay back the cost of the system. This is the case examined by EPRI and, if the slow acceptance of solar water heaters with their potentially much faster payback time is any indication, the least likely to succeed.

2. The utility itself may opt to invest in a large number of small solar cell arrays mounted on blocks of its customers'

homes, saving a substantial part of the balance of systems costs by simply renting the roof space through some equitable arrangement. This option should be particularly attractive to small rural electric coops that are member-owned and must currently purchase all their power from large utilities. Having the members participate in producing their own power is much more logical than attempting to invest in a part of a nuclear power plant that might never get built.

3. Independent third-party investors could put up the capital for systems on an entire subdivision under construction. The homeowner is compensated for the use of the roof and the power is sold directly to the utility. Such a financial package would work particularly well with a solar module such as the SunWatt panels with water-cooled cells. Since these hybrids make hot water and electricity at the same time, the homeowner would rent the array for the hot water output while the third-party investor receives all the revenue from the sale of the electricity.

Any of these distributed systems would require multiple inverters and controls to ensure safe and reliable operation, but the technical problems are well within the capability of our electronics industry. The main impediments will be political and social: people will have to get together and boldly work out the plans for our future use of renewable energy to furnish the electric power so necessary to the good life we desire.

RECOMMENDED READINGS

DeMeo, E. A., and R. W. Taylor. "Solar Photovoltaic Power Systems: An Electric Utility R&D Perspective," *Science* 224(4646): 245-251.

Proceedings of the IEEE Photovoltaic Specialists Conference, May 1, 1984 (New York: IEEE).

Chapter 9
The Future of Photovoltaics

The future of photovoltaics looks bright. An industry has been formed and myriad uses for photovoltaic devices have been discovered and developed. As the prices of cells and modules drop, the applications for which solar cells are the best choice will greatly increase and the market will grow exponentially. The prices will decrease, over the long run, not just because the expanded market allows for more efficient production but because research and development will produce less expensive photovoltaic devices. The speed at which innovation and development of production capacity will occur is sensitive to decisions made by government agencies, and influenced by the acquisition of small innovative companies by large oil companies. Obstacles will arise as inflationary pressures temporarily raise the price of solar cell modules and the lack of capital keeps small companies from implementing their automation and expansion plans. But these plans will become reality, new types of solar cells will become available, and we will be using them increasingly in the near future.

Since the space program began in the late 1950s, the U.S. government has been the world's major purchaser of solar cells. Most photovoltaic research and development has been financed by the government; in fact, the present-day photovoltaic cell would probably not exist except for this commitment. The expenditure of tax money on solar cell research has not been a matter of altruism, but rather the purchasing of a technology that was, and will continue to be, vitally important to our space program, and one that is fast becoming important in meeting our modern military needs. Our commercial satellite communications network would not exist without this commitment to photovoltaic research. In the future the government will use ever larger quantities of solar cells for both space and terrestrial applications, as photovoltaics proves to be, in many situations, the most economical, reliable and, quite often, secure way to generate electrical energy.

Since the formation of the U.S. photovoltaic program in 1974, the government has clearly defined the goals for the emerging photovoltaic industry. The Department of Energy established guidelines for the development of terrestrial solar cells that, if met, would result in solar cells replacing 1 quad of fossil or nuclear fuel by the year 2000 (1 quad is 10^{15} Btu or 3×10^{11} kWh). The DOE hopes that lowering the cost of solar cells will stimulate the market.

Figure 9.1 shows government projections for solar cell costs. The $2.00 per peak watt predicted for 1983 was calculated as the price necessary to attract users of electricity in remote areas. While that goal was not met, these users proved interested in solar cells, even at present prices.

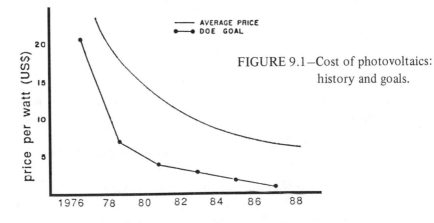

FIGURE 9.1—Cost of photovoltaics: history and goals.

A great deal has been written about U.S. government involvement in photovoltaics and about the relationship between price and market penetration. (See the Recommended Readings following this chapter.) In light of the Reagan administration's different philosophy in regard to solar energy, it is unclear, even with an expanded time schedule, whether the government is still interested in meeting the goals shown in Figure 9.1.

The Solar Energy Research Institute (SERI) in Golden, Colorado, had a very active photovoltaic research program. Although greatly constrained by budget cuts, an abbreviated SERI continues, studying those systems which are too new or poorly developed to be of interest to profit-oriented business, as well as implementing basic research into the physics and chemistry of photovoltaics. An original function of SERI, that of information source, has been drastically curtailed, and the regional solar centers formed with SERI have been phased out.

The federal government and many states had established tax incentives to encourage individuals to invest in home energy conservation projects and to install solar and wind energy systems. The federal tax credits, included in the "oil windfall profits" tax act, allowed 40% of the total cost of certain kinds of solar installations to be deducted from income taxes. The end of this tax credit has slowed the expansion of the PV industry, but the lower price of PV arrays has helped offset this loss of tax credits for many projects, particularly in remote areas. Many states still have energy credits. Contact your state energy office to find out about your state's solar tax incentives, or talk to your local state representative. If your state has no solar incentives, ask why.
contact your state solar office, if there is one, or talk to your local state representative. If your state has no solar incentives, ask why. When President Reagan took office, he announced his intention to rescind the federal tax credits in the interest of "balancing the budget." The current federal tax credits are due to expire in 1985, and it is unlikely they will be extended in their present form.

FOREIGN SOLAR CELL ACTIVITY

The photovoltaic industry is international in every sense of the word. Not only are solar arrays used in every country, but major contributions to the development of photovoltaics are made by private companies, universities and government laboratories worldwide. To list all the programs would be impossible. Some important new developments, however, should be mentioned.

Japan

Japan has the greatest commitment to research and development of solar cells next to the United States. The goal of the Japanese government's "Sunshine Project" is to reduce Japan's dependence on imported fuels by the extensive use of solar energy. The close cooperation between government, academia and private industry has already led to the rapid development of new amorphous silicon solar cells. The recently announced a-silicon carbide–silicon cell is an outstanding example of this technology. At the present time, the only commercially available Japanese solar cells are small ones used for pocket calculators and digital watches, but as explained in the previous chapter, Japanese companies should be selling amorphous silicon solar cell modules and arrays at extremely competitive prices very soon. Perhaps Japanese companies have not yet marketed conventional silicon power modules (which are certainly within their capabilities) because they have decided to put their resources into leap-frogging the current technology. Kyocera, for example, has recently started selling polycrystalline silicon solar modules in the United States. The cells are made using new crystal-growing techniques.

Western Europe

The photovoltaic cell was invented by a Frenchman, Edmond Becquerel, in 1839, and French scientists have never lost their interest in solar energy. In past years, the French government had been less than excited about the large-scale development of photovoltaic cells (possibly because of their heavy commitment to nuclear power). Since the advent of the Mitterand government, however, the emphasis is shifting toward renewable energies and new R&D efforts in photovoltaics. Individual French scientists and companies have been continuously in the forefront of solar cell research, working on low-cost systems. French companies seem to get into the photovoltaics business by buying small American manufacturers rather than setting up their own facilities. This trend may be changing since recently two French oil companies dropped their American subsidiaries, while continuing their own research.

The German government has begun to develop and encourage the use of photovoltaics, and a number of German university scientists, including Bloss in Stuttgart and Goetzberger at the Fraunhofer Institute, are exploring new photovoltaic systems. Germany's commercial development of photovoltaics has been largely in the area of furnishing raw silicon materials for solar cell manufacture. The Siemens process for purifying silicon for the semiconductor industry has become the standard method. Telefunken pioneered the method of casting cubes of polycrystalline, solar-grade silicon. These companies are expanding their production facilities (Siemens is greatly expanding its American facilities) and developing new lower-cost processes, and definitely anticipate capturing a large share of the growing photovoltaics market. They envision that their greater market will be in underdeveloped countries, rather than domestic.

Scientists in other Western European countries are closely following the developments in photovoltaic research but, except for Italy, most seem not to expect much impact on their energy needs. Italy's combination of Mediterranean sunshine and high cost of electric generation make solar cells an ideal way to implement their energy independence goals. The Italian photovoltaics program now underway includes the development of new photovoltaic devices as well as the manufacture of solar cells and modules.

Underdeveloped Countries

Solar cells, because they are currently so useful for powering small electric systems in remote locations, could make a great impact on countries that do not yet have central utility grids. These countries are taking a great interest in the progress of photovoltaics. Moreover, the World Bank and other international organizations are concerned with the tendency of underdeveloped countries to spend an ever greater part of their gross national product and to go deeper into debt to finance their energy needs.

Underdeveloped countries can get involved in using solar cells in two ways. The easiest is to simply import and use solar arrays made elsewhere; this market is becoming a major one for American photovoltaic manufacturers. Solar irrigation pumps have been developed, notably by the French for use in North Africa, and solar-powered television sets are enabling remote villages to receive educational programs from satellite stations. For dispersed uses of electric power, solar cells (and wind generators in the appropriate climates) make much more sense, both from an economic and a cultural point of view, than dependence on large centralized nuclear- or fossil fuel-powered electric plants. The cost of running power lines through the wilderness and the problems of maintaining such systems would put an intolerable burden on the resources of most developing countries. So solar cells will be used widely in the Third World long before they are cheap enough to be practical for the American homeowner.

The second way a country can benefit from the development of photovoltaic cells is to participate in their development and manufacture, either for internal consumption or for export. A number of countries—including Mexico, China and Yugoslavia—have initiated photovoltaic programs to meet their internal needs and have established pilot plants to make silicon solar cells.

It is important in a rapidly changing field like photovoltaics that a country not get locked into an obsolete technology and that development plans remain flexible enough to take advantage of breakthroughs in low-cost cells. Scientists and engineers in a country with a healthy research program can actually be in the forefront of such a new field and can make significant contributions, even if the program is small.

Many of the new types of photovoltaic cells now being developed, such as amorphous semiconductor cells and compound heterojunction devices, use simpler manufacturing techniques than those necessary for single-crystal cells. So it should be possible to set up relatively small manufacturing plants in developing countries to fill the energy needs of an area without disrupting the country's economy or requiring a large cadre of highly skilled foreign technicians. Companies already engaged in the assembly of automobiles or the manufacture of consumer goods will soon be able to tool up for the manufacture and assembly of solar devices, such as hybrid photovoltaics/solar heating modules. Unlike so many fruits of modern civilization, which have been paid for by the social disruption and economic dependence of developing countries, solar energy promises to be a means through which such nations can truly develop independently.

At least three American solar cell companies have announced joint ventures in developing countries to produce both solar cells and modules. Spire Corporation has several customers for their pre-packaged module assembly plant and Chronar is building an amorphous silicon cell production facility in Egypt. Recently SunWatt announced agreements with companies in Botswana and the Philippines to manufacture photovoltaic modules.

THE SOLAR ELECTRIC HOME

Thousands of homes are already powered by photovoltaic arrays. Most of these systems are small ones installed on remote dwellings by homesteaders who wish to power a few lights, a radio or television, but who are far from power lines. Some have connected their systems to the utility grid and engage in buy/sell arrangements with a power company. But increasingly we are seeing fairly elaborate systems installed by people who want a fully equipped household but choose not to deal with a power company.

With recent advances in inverter design and the ready availability of efficient low-voltage lighting and appliances, a sub-industry of photovoltaic installers and balance-of-systems specialists has emerged. These experts can furnish reliable, electrical household systems that are in every way the equivalent of the conventional grid-connected home. In the event of storm-caused power outages, especially in rural areas, the reliability of the stand-alone photovoltaics powered home can actually be greater than that of one connected to a utility grid.

The economics of the stand-alone home power system is becoming increasingly favorable. When you compare the system cost after applicable tax credits with the cost of conventional wiring and the power company's charges to run utility lines to the homesite, the price per kilowatt hour is about the same. If the site is some distance from the utility source, *e.g.*, a mile, a stand-alone system

FIGURE 9.2—The occupants of this completely energy-independent home in Southern Indiana use two SunWatt photovoltaic/hot water hybrid modules to furnish all of their electrical and hot water needs.   (Skyheat photo)

FIGURE 9.3—The PV-powered lawn mower predicted by *Business Week* in the 1950s is now a reality.

(Skyheat photo)

will actually cost less than the $10,000 or so charged by the power company to extend lines.

Sizing, designing and installing the system is more elaborate than wiring a conventional home, but not too complicated. *The New Solar Electric Home* gives detailed instructions. When the weather data for most parts of the country are compared to the electrical needs, one finds a bad short-fall in the winter, necessitating a large photovoltaic array to handle these few bad months. Combining solar cells with a wind generator can, in many places, produce a reasonably constant supply of power and allow a reduction in the size of the battery bank needed to store the power. **compuSOLAR** computer programs for sizing systems include an excellent method for determining the proper combination of sun- and wind-powered systems.

The person contemplating a self-sufficient home must reduce power consumption to a fraction of that used by the typical American homeowner. This can now be done with little change in lifestyle. Very efficient alternatives exist for most home appliances, and quite often the quality of life will be enhanced by some of the changes. One advantage of a passive solar home, for example, is the delightful living space created, and with a set of hybrid photovoltaic/hot water modules, all your hot water needs, including luxuries, can be accommodated.

In the future, more and more people will choose the path of energy independence, and not simply to save money. Consumer resentment at being asked to pay for the mistakes of utility executives will drive more people to disconnect from the utility grid. Financing cancelled nuclear power plants, and the growing awareness of the dangers of considering them in the first place, will lead to the realization that viable alternatives exist in the form of reliable renewable energy systems.

SOLAR CELLS IN SPACE

From the beginning of the space program, solar cells have played an important role in powering satellites. The original Vanguard satellite's one-watt solar array performed flawlessly. Since then the space program and photovoltaic cells have become soph-

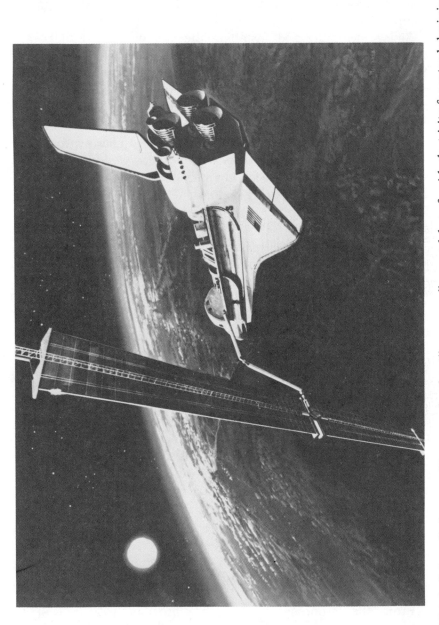

FIGURE 9.4—Solar cells in space. The space shuttle will carry a roll-up module to furnish electricity for extended missions.

isticated to the point where millikilowatt arrays are routinely used on communication satellites. The high-efficiency, blue-sensitive silicon cell that has become the standard design for commercial terrestrial cells was originally developed by Joseph Lindmayer for space applications. The importance of quality control in manufacture and the consequent reputation for reliability that photovoltaic modules earned was part of the space program philosophy. Reliability will remain an important factor in the ready acceptance of solar cells for remote terrestrial applications, once the price becomes low enough to make them competitive with other small power sources.

With the advent of the space shuttle, the number of solar cells used in space is growing enormously. Researchers are working on new cell designs that will be even more reliable, more efficient, easier to assemble, and less expensive than those space cells now used. The space shuttle will carry onboard a large array that can be unrolled in space to extend the mission time long past that possible with the hydrogen fuel cells now used. Figure 9.4 shows an artist's conception of an even larger array designed to be left in space and used as a power station for the space shuttle.

The Solar Power Satellite

One idea proposed for large-scale solar generation of electric power is the solar power satellite. First proposed by Peter Glaser, the solar power satellite is envisioned as a multimegawatt photovoltaic array parked in a synchronous orbit 22,000 miles out in space beaming the photogenerated power back to earth in the form of microwaves. This energy would be picked up by antenna farms, miles in diameter, placed in remote areas or floating offshore in the oceans. A network of these satellites could furnish all the electricity our planet needs more cheaply and with much less environmental disruption than the equivalent number of nuclear or coal-fired plants. For proponents of the status quo, the solar power satellite system has the additional advantage of being a centralized source of electricity, feeding into the existing electric grid, that could be owned and controlled by the present utility companies.

There are a number of safety and environmental questions to be answered concerning the effects of the enormous amount of microwave energy that would be focused onto selected spots on earth—although the plan is to spread the beam over such a large area that its intensity is diluted to the point that no known harm would be done. Carefully designed feedback controls would ensure the correct aim of the beam from such an enormous distance, and safety controls would quickly shut down the satellite transmitter in the event that the beam did wander off the target receiver. The design of the safety control system would actually be simpler than that of a nuclear power plant, and the consequences of a failure would most likely be considerably less. Before such a plan is considered, however, intensive research should be undertaken to study the long-term effects of low-level microwave radiation; accidents are inevitable and portions of the population could occasionally be bathed in microwaves.

Another question is an economic one. The construction of the network as envisioned requires the building of possibly dozens of enormous photovoltaic power satellites composed of tons of materials and millions of high-efficiency solar cells. The capital requirements are so large that only the most industrialized nations or a consortium of many giant corporations could even consider the project. The great number of twice weekly space shuttle trips necessary to deliver parts to synchronous orbit would consume an enormous amount of petroleum and could cause widespread environmental effects. After a three-year preliminary feasibility and design study, the federal government has decided to suspend funds for further work on the solar power satellite because of these economic and environmental problems.

A Proposed Alternative

Among other advantages, it may be safer and cheaper in the long run to start a space colony with solar cell manufacturing facilities and utilize raw materials from the moon or asteroids to construct the power satellites. Such a proposal has been worked out in some detail by O'Neal of Princeton and others. Manufacturing solar cells in space may turn out to be a very profitable enterprise. Many of the problems that plague the manufacture of inexpensive, high-quality silicon cells may be simplified where a gravity-free environment and a high-quality vacuum are so easy to obtain.

Certainly the space colony will use solar energy to meet all its energy needs, and a plentiful supply of energy will also help the fledgling space solar cell industry. Making solar cells is presently an energy-intensive business and calculations have been made of the growth rate of a manufacturing facility (called the "solar breeder") if a given percentage of the output cells is dedicated to furnishing the power needs of the facility. The only external supply of energy is that needed to construct the initial nucleus of the factory; all other energy needs come from the dedicated portion of the output. In space, where AMO (1.4 kW/m^2) of sunlight is continuously available, the growth rate for the solar breeder could be fast indeed.

Rather than build a solar power satellite, it might be more profitable yet for the space colony to ship manufactured solar modules to earth to be sold directly to homeowners as photovoltaic roof shingles. This approach would eliminate the cost and complexity involved in converting solar electricity to microwaves and constructing the receiving antennas and power conditioning equipment needed to convert the microwaves back to electricity. Instead of continuously beaming energy, the space colony would only have to ship a bundle of shingles once; the downhill trip on the space shuttle could be made profitable, since otherwise it would quite often return to earth empty.

The sum total of the area required for the proposed power satellite's receiving antennas is a significant fraction of the roof areas of the customers of the electricity generated and actually not that much less than the area required by the terrestrial photovoltaic arrays needed to produce the equivalent amount of electricity.

The problems associated with this scenario are common to all decentralized electricity-generating networks. The perennial problems of balancing the available energy supply and consumption at any particular moment would be minimized by increasing the size of the utility grid and the incorporation of centralized short-term storage so that the decentralized producer network could actually be made more reliable than our present patchwork utility grid. The social and political problems would be much more difficult to solve. Our society would have to change from one whose consumers depend completely on a few large corporations and government agencies for their energy needs to a society whose consumers are self-sufficient but cooperating producers of energy. Some large institutions are already resisting the tentative steps in this direction. However, these steps are being taken by pioneering individuals and groups who use the sun, wind and water to generate more electric power than they need in an attempt to sell the excess to local utilities. So far these attempts have been thwarted—utilities have filed suits to reverse the present law—but these pioneers will ultimately succeed. The price of solar cells will come down, we will eventually have decentralized power grids, and photovoltaics will become the most practical way to furnish the energy our civilization requires.

RECOMMENDED READINGS

Commission of the European Communities. *Third E.C. Photovoltaic Solar Energy Conference* (Dordrecht, Holland: D. Reidel Publishing Co., 1981).

Cook, Stephen. Information packet. **compuSOLAR,** Gum Springs Road, Jasper, AR 72641.

Davidson, Joel. *The New Solar Electric Home: The Photovoltaics How-To Handbook* (Ann Arbor, MI: **aatec publications,** 1987).

Maycock, Paul, and Edward Stirewalt. *Photovoltaics: Sunlight to Electricity in One Step* (Andover, MA: Brick House Publishing Co., 1981).

Photovoltaic Venture Analysis. Final Report. Solar Energy Research Institute, SERI/TR-52-040 (1978).

APPENDIX A

MANUFACTURERS OF SOLAR CELLS AND MODULES

(Most of the manufacturers listed below sell modules to the
general public, but only a few sell individual cells.)

Alpha Solarco
600 Vine Street
Cincinnati, OH 45202
513 621 1243

Applied Solar Energy Corporation
15751 East Don Julian Road
City of Industry, CA 91746
818 968 6581

ARCO Solar, Inc.
PO Box 6032
Camarillo, CA 93010
805 388 6389

Chronar Corporation
330 Basin Road
Lawrenceville, NJ 08648
609 799 1348

Energy Conversion Devices
1675 West Maple Road
Troy, MI 48084
313 280 1900

Entech
PO Box 612
DFW Airport, TX 75261
214 272 0515

Heliopower
One Centennial Plaza
Piscataway, NJ 08854
201 980 0707

Kyocera International Inc.
8611 Balboa Avenue
San Diego, CA 92123
619 576 2647

Photron
77 West Commercial Street
Willits, CA 95490
707 459 3211

Silicon Sensors, Inc.
Highway 18 East
Dodgeville, WI 53533
608 935 2707

Solarex Corporation
1335 Piccard Drive
Rockville, MD 20850
301 948 0202

Solec International, Inc.
12533 Chadron Avenue
Hawthorne, CA 90250
213 970 0065

Spectrolab, Inc.
12500 Gladstone Avenue
Sylmar, CA 91342
818 365 4611

Spire Corporation
Patriots Park
Bedford, MA 01730
617 275 6000

SunWatt Corporation
RFD Box 751
Addison, ME 04606
207 497 2204

Tideland Signal Corp
PO Box 52430
Houston, TX 77052
713 681 6101

APPENDIX B

CURRENT CARRYING CAPACITY OF COPPER WIRE

The ratings in the following tabulations are those permitted by the National Electrical Code for flexible cords and for interior wiring of houses, hotels, office buildings, industrial plants, and other buildings.

The values are for copper wire. For aluminum wire the allowable carrying capacities shall be taken as 84% of those given in the table for the respective sizes of copper wire with the same kinds of covering.

Size A.W.G.	Area Circular (mils)	Diameter of Solid Wires (mils)	Rubber Insulation (amps)	Varnished Cambric Insulation (amps)	Other Insulations and Bare Conductors (amps)
24	404	20.1	—	—	1.5
22	642	25.3	—	—	2.5
20	1,022	32.0	—	—	4
18	1,624	40.3	3*	—	6**
16	2,583	50.8	6*	—	10**
14	4,107	64.1	15	18	20
12	6,530	80.8	20	25	30
10	10,380	101.9	25	30	35
8	16,510	128.5	35	40	50
6	26,250	162.0	50	60	70
5	33,100	181.9	55	65	80
4	41,740	204.3	70	85	90
3	52,630	229.4	80	95	100
2	66,370	257.6	90	110	125

Note: 1 mil = 0.001 inch.

*The allowable carrying capacities of No. 18 and 16 are 5 and 7 amperes, respectively, when in flexible cord.

**The allowable carrying capacities of No. 18 and 16 are 10 and 15 amperes, respectively, when in cords for portable heaters. Types AFS, AFSI, HC, HPD and HSJ.

177

APPENDIX C

CONVERSION FACTORS

To Change	Into	Multiply by
BTU	cal	252
BTU	joules	1,055
BTU	kcal	0.252
BTU	kWh	2.93×10^{-4}
BTU ft^{-2}	langleys (cal cm^{-2})	0.271
cal	BTU	3.97×10^{-5}
cal	ft-lb	3.09
cal	joules	4.184
cal	kcal	0.001
cal min^{-1}	watts	0.0698
cm	inches	0.394
cc or cm^3	in.^{3}	0.0610
ft^3	liters	28.3
in.^{3}	cc or cm^3	16.4
ft	m	0.305
ft-lb	cal	0.324
ft-lb	joules	1.36
ft-lb	kg-m	0.138
ft-lb	kWh	3.77×10^{-7}
gal	liters	3.79
hp	kW	0.745
inches	cm	2.54
joules	BTU	9.48×10^{-4}
joules	cal	0.239
joules	ft-lb	0.738
kcal	BTU	3.97
kcal	cal	1,000
kcal min^{-1}	kW	0.0698
kg-m	ft-lb	7.23
kg	lb	2.20
kW	hp	1.34
kWh	BTU	3,413

kWh	ft-lb	2.66×10^6
kW	kcal min^{-1}	14.3
langleys (cal cm^2)	BTU ft^{-2}	3.69
langleys min^{-1} (cal cm^{-2} min^{-1})	watts cm^{-2}	0.0698
liters	gal	0.264
liters	qt	1.06
m	ft	3.28
lb	kg	0.454
qt	liters	0.946
cm^2	ft^2	0.00108
cm^2	in.^{2}	0.155
ft^2	m^2	0.0929
m^2	ft^2	10.8
watts cm^{-2}	langleys min^{-1} (cal cm^2)	14.3

GLOSSARY

AC—Alternating current; the electric current which reverses its direction of flow. 60 cycles per second is the standard current used by utilities in the U.S.

acceptance angle—The total range of sun positions from which sunlight can be collected by a system.

acceptor levels—Levels capable of accepting an electron from the valence band.

AH—*see* **ampere-hours**

Air Mass 0 (AM0)—The amount of sunlight falling on a surface in outer space just outside the earth's atmosphere (1.4 kW/m^2).

Air Mass 1 (AM1)—The amount of sunlight falling on the earth at sea level when the sun is shining straight down through a dry clean atmosphere. (A close approximation is the Sahara Desert at high noon.) The sunlight intensity is very close to 1 kilowatt per square meter (1 kW/m^2).

Air Mass 2 (AM2)—A closer approximation to usual sunlight conditions; may be simulated by an ELH projector bulb. The illumination is 800 W/m^2.

alternating current—*see* **AC**

amorphous semiconductors—*see* **semiconductors, amorphous**

amorphous silicon—A form of silicon with no long-range crystalline order used to make solar cells.

ampere-hours (AH)—A current of one ampere running for one hour.

balance-of-systems (BOS)—All the equipment needed to make a complete working photovoltaic power system, except the solar cell modules.

band, conduction—*see* **conduction band**

band, valence—*see* **valence band**

band gap energy—The amount of energy needed to raise an electron from the top of the valence band to the bottom of the conduction band.

band model—The quantum mechanical model of solids that explains the behavior of semiconductors.

black-body radiation—*see* **radiation, black-body**

battery, marine—A deep-discharge battery used on boats; capable of discharging small amounts of electricity over a long period of time.

battery, motive-power—A large-capacity deep-discharge battery designed for long life when used in electric vehicles.

battery, secondary—*see* **battery, storage**

battery, stationary—For use in emergency standby power systems, a battery with long life but poor deep-discharge capabilities.

battery, storage—A secondary battery; rechargeable electric storage unit that operates on the principle of changing electrical energy into chemical energy by means of a reversible chemical reaction. The lead-acid automobile battery is the most familiar of this type.

blocking diode—A device that prevents the current from running backward through an array, thereby draining the storage battery.

built-in potential—The electrical potential that develops across the junction when two dissimilar solids are joined. The open-circuit voltage of a photovoltaic cell will approach but always be less than this potential.

bulk recombination—*see* **recombination, bulk**

carriers, majority—The carrier most present in a doped semiconductor: holes in p-type, electrons in n-type.

carriers, minority—The carrier least present in a doped semiconductor: electrons in p-type, holes in n-type.

cascade cells—*see* **tandem cells**

cell capacity—Expressed in ampere-hours, the total amount of electricity that can be drawn from a fully charged battery until it is discharged to a specific voltage.

composite cell structure—A device consisting of two solar cells built atop one another as one unit. The top cell absorbs short wavelength light and allows the longer wavelengths to pass through to illuminate the lower cell. A tunnel junction separates the cells. *Also see* **tunnel junction.**

concentration ratio—The ratio between the area of clear aperture (opening through which sunlight enters) and the area of the illuminated cell.

conduction band—The upper, usually empty band in a semiconductor. Electrons with this energy are free to move throughout the solid.

conductivity, intrinsic—The electrical conductivity of an undoped semiconductor material. This conductivity, which is extremely small in solar cell materials, is produced by the direct thermal activation of electrons from the valence band to the conduction band. It is very temperature-dependent.

current–voltage curve—I-V curve; plots current on the vertical axis versus the voltage on the horizontal axis.

Czochralski process—Method of growing a single crystal by pulling a solidifying crystal from a melt.

DC—Direct current; electric current that always flows in the same direction—positive to negative. Photovoltaic cells and batteries are all DC devices.

deep-discharge cycles—Cycles in which a battery is nearly completely discharged.

depletion layer—The layer between the n and p layers at the junction where there are essentially no carriers.

diffusion furnace—Furnace used to make junctions in semiconductors by diffusing dopant atoms into the surface of the material.

direct current—*see* **DC**

direct band gap semiconductor—*see* **semiconductor, direct band gap**

donor level—The level that donates conduction electrons to the system.

doping—The deliberate addition of a known impurity (**dopant**) to a pure semiconductor to produce the desired electric properties.

electrolyte—A liquid conductor of electricity.

electron affinity—The energy difference between the bottom of the conduction band and vacuum zero.

electronic-grade silicon—*see* **silicon, electronic-grade**

energy levels—The energy represented by an electron in the band model of a substance.

extrinsic semiconductors—*see* **semiconductors, extrinsic**

fermi level—Energy level at which the probability of finding an electron is one-half. In a metal, the fermi level is very near the top of the filled levels in the partially filled valence band. In a semiconductor, the fermi level is in the band gap.

fill factor (ff)—The actual maximum power divided by the hypothetical "power" obtained by multiplying the open-circuit voltage by the short-circuit current.

fingers—*see* **front contact fingers**

fresnel lens—A segmented lens, usually molded of plastic, used to concentrate sunlight onto a receiver.

front contact fingers—The thin, closely spaced lines of the front electrode that pick up the current from the semiconductor but allow light to pass between them into the solar cell.

heterojunction—A solar cell in which the junction is between two different semiconductors.

holes—An energy level that could be occupied by an electron, but currently is not. Holes act like charged particles, with energy and momentum, and are capable of carrying an electric current.

homojunction—A solar cell made from a single semiconductor; the junction is formed between n- and p-type doped layers.

hybrid system—A system that produces both usable heat as well as electricity.

hydrogen economy—A system in which hydrogen is substituted for fossil fuels. The hydrogen is basically a means of transporting and storing renewable energy.

indium oxide—A wide band gap semiconductor that can be heavily doped with tin to make a highly conductive transparent thin film. Often used as a front contact or one component of a heterojunction solar cell.

infrared wavelengths—Wavelengths longer than 700 nm; they cannot be seen, but are felt as "heat."

intrinsic conductivity—*see* **conductivity, intrinsic**

intrinsic semiconductors—*see* **semiconductors, intrinsic**

inverter—A device that converts DC to AC.

inverter, synchronous—A device that converts DC to AC in synchronization with the power line. Excess power is fed back into the utility grid.

ion implantation—A method of doping semiconductors by striking the surface with a beam of high-energy ions.

I_{SC} —*see* **short-circuit current**

ITO—Indium oxide doped with tin oxide.

I-V curve—*see* **current–voltage curve**

junction diode—A semiconductor device with a junction and a built-in potential that passes current better in one direction than the other. All solar cells are junction diodes.

junction, liquid—A junction in which one side is a liquid electrolyte.

junction, metal-insulator semiconductor (MIS solar cell)—A junction containing a thin (20 Å) insulating layer between the metal and the semiconductor.

junction, metal-to-semiconductor (MS junction)—A junction produced when a high work function (noble) metal is placed on a n-type semiconductor, or a low work function metal on a p-type semiconductor. (Schottky barrier junction.)

junction, tunnel—A special junction between two solar cells in a composite cell structure. Carriers cross the junction by quantum mechanical "tunneling." *See* **tunneling** and **composite cell.**

majority carriers—*see* **carriers, majority**

metallurgical-grade silicon—*see* **silicon, metallurgical-grade**

minority carriers—*see* **carriers, minority**

MS junction—*see* **junction, metal-to-semiconductor**

n-type semiconductors—*see* **semiconductors, n-type**

ohmic contacts—Contacts that do not impede the flow of current into or out of the semiconductor.

open-circuit voltage—An equilibrium voltage, reached when the number of carriers drifting back across the junction is equal to the number being generated by the incoming light.

parallel—In module construction; to increase current output, cells are wired with the back contact of one cell connected to the back contact of the next. The total current is the sum of the individual current outputs of the cells, but the total voltage is the same as the voltage of a single cell. Cells are usually wired in series to form an array and arrays are wired in parallel to obtain desired current.

passivate—To chemically react a substance with the surface of a solid to tie up or remove the reactive atoms on the surface. For example, the air oxidation of a fresh aluminum surface to form a thin layer of aluminum oxide.

polycrystalline solar cells—Cells in which more than one crystal (and crystal boundaries) are present in an individual cell.

power tower—A device that generates electric power from sunlight, consisting of a field of small mirrors tracking the sun to focus the light onto a tower-mounted boiler. The steam produced runs a conventional turbine generator.

p-type semiconductors—*see* **semiconductors, p-type**

radiation, black-body—Radiation emitted by the sun; it is composed of different wavelengths—some visible.

recombination center—A point, often an impurity or defect, where a hole–electron pair created by the absorbed light can recombine before the electron can pass through the external circuit; acts like an internal short circuit.

recombination, bulk—Occurs when the recombination centers are crystal defects or impurities in the bulk of the semiconductor.

recombination, surface—Occurs when recombination centers are surface impurities or surface states. Sometimes caused by damage to the surface. These can often be removed by passivation and/or chemical etching.

recombination, junction—Occurs when the recombination centers are at the junction or in the depletion layers. In heterojunctions it is often caused by crystal lattice mismatch between the two semiconductors.

rectifier—A device that passes current in one direction only.

refractive index—A measure of the amount of bending or "refraction" light undergoes when passing into or out of a substance.

Schottky barrier junction—*see* **junction, metal-to-semiconductor**

secondary battery—*see* **battery, storage**

self-discharge rate—The rate at which a battery will discharge on standing; affected by temperature and battery design.

semiconductors, amorphous—Semiconductor with no long-range crystal order.

semiconductors, direct band gap—The light absorbed can cause electrons to jump directly from the top of the valence band to the bottom of the conduction band. These semiconductors absorb light very strongly and can be used as very thin films.

semiconductors, extrinsic—The product of doping a pure semiconductor.

semiconductors, indirect band gap—When momentum conditions forbid the direct jumping of electrons; a more complex transition is required and light is absorbed less strongly, requiring thicker solar cells. Silicon is an indirect band gap semiconductor.

semiconductors, intrinsic—An undoped semiconductor. *Also see* conductivity, intrinsic.

semiconductors, n-type—Semiconductor in which negative electrons carry the current; produced by doping an intrinsic semiconductor with an electron donor impurity (phosphorus in silicon).

semiconductor, p-type—Semiconductor in which positive holes carry the current; produced by doping an intrinsic semiconductor with an electron acceptor impurity (boron in silicon).

series—In array construction; connecting cells by joining the back contact of one cell to the front contact of the next cell to obtain a higher voltage.

shingling—In array construction; connecting cells by overlapping the front edge of one cell with the back edge of the next, similar to roof shingles, and soldering the edges together.

short-circuit current (I_{SC})—The maximum current a cell can deliver into a short circuit; directly proportional to the area of the cell and the light intensity.

silicon, electronic-grade—Highly purified silicon needed for the manufacture of semiconductor devices (semiconductor-grade). Very expensive and in short supply.

silicon, metallurgical-grade—99.8% pure silicon suitable for most industrial uses. Relatively inexpensive.

silicon, solar-grade—Intermediate-grade silicon proposed for the manufacture of solar cells. Should be much less expensive than electronic-grade.

solar array—A set of modules assembled for a specific application; may consist of modules in series for increased voltage or in parallel for increased current, or a combination of both. *Also see* solar module.

solar cell—A device that converts sunlight directly into electricity.

solar-grade silicon—*see* silicon, solar-grade

solar module—A series string of 32 to 36 cells, producing an open-circuit voltage in bright sunlight of about 18 volts, or 16 volts when producing maximum power. Total current output of a series string is the same as a single cell. *Also see* solar array.

solar power farms–Large centralized photovoltaic array systems that generate electricity to power the utility grid.

sputtering–A method of depositing thin films utilizing a low-pressure gas discharge, either DC or radio-frequency, to knock atoms or molecules off an electrode onto the substrate to be coated.

sulfation–A condition which afflicts unused and discharged batteries; large crystals of lead sulfate grow on the plate, instead of the usual tiny crystals, making the battery extremely difficult to recharge.

surface recombination–*see* **recombination, surface**

tandem solar cells–Two photovoltaic cells are constructed on top of each other such that the light passes through the wide band gap cell to reach the narrow band gap cell. These cells very efficiently utilize the sunlight giving a greater output for a given area. Also called **cascade cells.**

tin oxide–A wide band gap semiconductor similar to indium oxide; used in heterojunction solar cells or to make a transparent conductive film called NESA glass when deposited on glass.

total energy system–Hybrid system producing both usable heat and electricity.

tracking system, 2-axis–A mount capable of pivoting both daily and seasonally to follow the sun.

tracking system, 1-axis–A mount pointing in one axis only; reoriented seasonally by hand and used with linear concentrators or flat plates.

tunneling–Quantum mechanical concept whereby an electron is found on the opposite side of an insulating barrier without having passed through or around the barrier.

ultraviolet (UV) wavelengths–Wavelengths shorter than 400 nm; the energetic rays of the sun, invisible but responsible for suntans and sunburns. Our atmosphere filters out most UV rays.

vacuum evaporation–Method of depositing thin coatings of a substance by heating it in a vacuum system.

vacuum zero—The energy of an electron at rest in empty space; used as a reference level in energy band diagrams.

valence band—The band of energy levels occupied by the valence electrons in a solid; always below vacuum zero.

V_{OC} —*see* **open-circuit voltage**

voltage, open-circuit—*see* **open-circuit voltage**

voltage, short-circuit—*see* **short-circuit voltage**

Winston concentrator—A trough-type parabolic collector with one-axis tracking developed by Roland Winston.

work function—The energy difference between the fermi level and vacuum zero. The minimum amount of energy it takes to remove an electron from a substance into the vacuum.

zone refining—Method of purifying solid rods by means of melting narrow zones through the rods. These zones are slowly moved from one end of the rod to the other, sweeping out the impurities.

INDEX